Men, Masculinities and Disaster

T0136273

In the examination of gender as a driving force in disasters, too little attention has been paid to how women's or men's disaster experiences relate to the wider context of gender inequality, or how gender-just practice can help prevent disasters or address climate change at a structural level.

With a foreword from Kenneth Hewitt, an afterword from Raewyn Connell and contributions from renowned international experts, this book helps address the gap. It explores disasters in diverse environmental, hazard, political and cultural contexts through original research and theoretical reflection, building on the under-utilized orientation of critical men's studies. This body of thought, not previously applied in disaster contexts, explores how men gain, maintain and use power to assert control over women. Contributors examine men's lives on the "gendered terrain of disasters," considering how diverse forms of masculinities shape men's efforts to respond to and recover from disasters and other climate challenges. The book highlights both the high costs paid by many men in disasters and the consequences of dominant masculinity practices for women and marginalized men. It concludes by examining how disaster risk can be reduced through men's diverse efforts to challenge hierarchies around gender, sexuality, disability, age and culture.

Elaine Enarson is an independent scholar based in Colorado, USA.

Bob Pease is Professor of Social Work at the University of Tasmania, Australia.

Routledge Studies in Hazards, Disaster Risk and Climate Change

Series Editor: Ilan Kelman, Reader in Risk, Resilience and Global Health at the Institute for Risk and Disaster Reduction (IRDR) and the Institute for Global Health (IGH), University College London (UCL)

This series provides a forum for original and vibrant research. It offers contributions from each of these communities as well as innovative titles that examine the links between hazards, disasters and climate change, to bring these schools of thought closer together. This series promotes interdisciplinary scholarly work that is empirically and theoretically informed, with titles reflecting the wealth of research being undertaken in these diverse and exciting fields.

Published:

Cultures and Disasters
Understanding cultural framings in disaster risk reduction
Edited by Fred Krüger, Greg Bankoff, Terry Cannon, Benedikt Orlowski and E. Lisa F. Schipper

Recovery from Disasters
Ian Davis and David Alexander

Men, Masculinities and Disaster
Edited by Elaine Enarson and Bob Pease

Men, Masculinities and Disaster

Edited by
Elaine Enarson and Bob Pease

Routledge
Taylor & Francis Group

LONDON AND NEW YORK

First published 2016
by Routledge

2 Park Square, Milton Park, Abingdon, Oxfordshire OX14 4RN
711 Third Avenue, New York, NY 10017

Routledge is an imprint of the Taylor & Francis Group, an informa business

First issued in paperback 2018

British Library Cataloguing-in-Publication Data
A catalogue record for this book is available from the British Library

Library of Congress Cataloging-in-Publication Data
Names: Enarson, Elaine Pitt, 1949– editor. | Pease, Bob, editor.
Title: Men, masculinities and disaster / edited by Elaine Enarson and Bob Pease.
Description: Abingdon, Oxon; New York, NY: Routledge is an imprint of the Taylor & Francis Group, an Informa Business, [2016] | Series: Routledge studies in hazards, disaster risk and climate change | Includes index.
Identifiers: LCCN 2016001722| ISBN 9781138934177 (hbk) |
ISBN 9781315678122 (ebk)
Subjects: LCSH: Disasters—Social aspects—Cross-cultural studies. | Disaster relief—Social aspects—Cross-cultural studies. | Masculinity—Cross-cultural studies. | Sex role—Cross-cultural studies.
Classification: LCC HV553 .M45 2016 | DDC 363.340811—dc23
LC record available at http://lccn.loc.gov/2016001722

ISBN: 978-1-138-93417-7 (hbk)
ISBN: 978-1-138-32460-2 (pbk)

Typeset in Times New Roman
by Book Now Ltd, London

Contents

Contributors

Duke W. Austin is an Assistant Professor of Sociology at California State University, East Bay and a Senior Fellow for the Urban Ethnography Project at Yale University. He researches the intersections of race, gender, immigration and disasters in the United States. His previous work explores these topics in the aftermath of Hurricane Katrina and in disaster preparation among San Francisco's community-based organizations. Professor Austin has also edited a volume for the *Annals of the American Academy of Political and Social Science* and published research on race as a caste-like system, a concept that was originally presented in the classic ethnographies *Caste and Class in a Southern Town* (1937) and *Deep South* (1941).

Dave Baigent was an operational firefighter for 31 years, an officer and a Fire Brigade Union representative in the London Fire Brigade. Dave took a first degree in sociology at Anglia Ruskin Cambridge and then a PhD on how firemen construct their masculinity. He used a profeminist autocritique as a means to interrogate firemen's identity and to help others to do the same. As a Principal Lecturer, Dave wrote and led the UK's first social science Public Service Degree, and undertook a variety of research and consultancy projects into the effects of masculinity in the fire service in the UK, Australia and in Sweden. In the last three years he has been advising the Swedish government on these issues and participating in education aimed at getting more women involved in the fire service.

Sarah Bradshaw is an Associate Professor at Middlesex University in London and has worked in gender and development for over 20 years, combining lecturing and researching with work with NGOs. While working in Nicaragua, she lived through Hurricane Mitch and was involved in the civil society initiatives to influence the national and international reconstruction program, including working on a number of research projects to explore the gendered impact. She has continued to work on both development and disasters, including publishing a recent book on the theme *Gender, Development and Disasters* (2013). She continues to combine academic and advocacy work. Most recently she has been working with the UN Sustainable Development Solutions Network seeking to influence the new Sustainable Development Goals.

Robin Cox is Professor and Program Head of the Disaster and Emergency Management programs at Royal Roads University in Victoria, British Columbia, where she also leads the Resilience by Design research lab (RbD). Robin's research brings together an interdisciplinary collaboration of researchers, students and practitioners interested in exploring individual and community resilience and the potential for sparking social change and creative innovation in the context of climate change and disaster risk reduction. Her current research projects focus on empowering youth as leaders and contributors to community resilience and sustainability through Creative Action Research and Social Innovation processes. These projects include research focused on engaging children and youth in an exploration of community resilience in post-flood communities in Southern Alberta, a participatory project on youths' post-disaster recovery in Canada and the US, and research focused on synthesizing existing knowledge on children's and youths' resilience in the context of energy resource production, climate change and the transition to low-carbon economies. Robin is also an active disaster psychosocial practitioner and responder.

Malathi de Alwis received her PhD in sociocultural anthropology from the University of Chicago and is currently affiliated with the Faculty of Graduate Studies, University of Colombo and the Open University, Colombo. She has also taught at the University of Chicago, New School for Social Research, New York, the International Women's University, Hannover and the University of Witwatersrand, Johannesburg. She has written extensively on nationalism, humanitarianism, maternalism, suffering and memorialization. She is the co-editor of *Tsunami in a Time of War: Aid, Activism and Reconstruction in Sri Lanka and Aceh* (with Eva-Lotta E Hedman; 2009), *Feminists Under Fire: Exchanges Across War Zones* (with Wenona Giles and Edith Klein; 2003) and *Embodied Violence: Communalising Women's Sexuality in South Asia* (with Kumari Jayawardena; 1996).

Dale Dominey-Howes is Associate Professor in the School of Geosciences at the University of Sydney. His expertise is in natural hazards, hazard, risk and vulnerability assessment, disaster and emergency management and interconnections between biophysical systems and the socioeconomic contexts in which disasters unfold. He has worked extensively with local, state and federal governments, universities and government agencies throughout Australasia, Asia, the Pacific and Europe on natural hazards and disaster risk reduction projects. His recent work includes developing a new integrated, multidisciplinary, multisectoral framework (a "coupled human-environment systems framework") for post-tsunami disaster impact assessment, published in *Nature*. In 2009 he was appointed by the UN to lead the largest post-tsunami disaster assessment team assembled in the South Pacific, and his framework was implemented by the United Nations in 2010.

Leith Dunn is Head of the Institute for Gender and Development Studies Mona Unit at the University of the West Indies (UWI) in Jamaica, and is a graduate of the London School of Economics and Political Science (PhD in sociology) and the UWI (BA Hons. and MSc). She has worked with national, regional and international development agencies including UNFPA and has published on gender mainstreaming, trade, child labor, tourism, human trafficking, HIV, governance, climate change and disaster management. Her publications include: the Commonwealth Foundation's *NGOs: Guidelines for Good Policy and Practice* (with Colin Ball; 1995); *From Kyoto to the Caribbean: Promoting Gender and Age Responses in Climate Change Policies in SIDS* (2013); *Enhancing Gender Visibility in Disaster Risk Management and Climate Change* (2009); and "The Gendered Dimensions of Environmental Justice: Caribbean Perspectives" (in Filomina Chioma-Steady's *Environmental Justice in the New Millennium*, 2009).

Elaine Enarson is an independent scholar based in Colorado. After receiving her PhD in sociology from the University of Oregon, she coordinated the Nevada Network Against Domestic Violence and the first women's studies program at the University of Nevada-Reno. Her US and global research has highlighted women's economic, cultural, community and political work in disasters, as well as gender issues in disaster housing and long-term recovery, women's human rights in disasters, and gender-based violence. A founding member of the Gender and Disaster Network, she speaks widely on the subject, writes on gender mainstreaming to reduce disaster risk, and co-edited three international books in the field. Her US-focused book *Women Confronting Natural Disaster: From Vulnerability to Resilience* was released in 2012. Currently, she teaches online courses to US and Canadian emergency management students, and is developing a website about disaster quilts.

Mathias Ericson currently works as an Associate Senior Lecturer and Researcher at the Department for Cultural Studies, Gothenburg University, Sweden. He holds a PhD in sociology with research interests in masculinity, risk and professions. His doctoral thesis, "Up Close: Masculinity, Intimacy and Community in Firefighters' Work Teams", is an ethnographic study of homosocial practices among male firefighters. He has worked with research projects on gender implications of educational restructuring within the firefighter profession and the shift from reactive to proactive modes within the rescue service. His current research interest concerns masculinity, vulnerability and necropolitics in the case of the broadened spectrum of mortality and risk calculation in rescue service work.

Christine Eriksen is a Research Fellow at the University of Wollongong, Australia. She specializes in social dimensions of disaster resilience. A major part of this work focuses on the culturally and historically distinct gender relations that underpin bushfire vulnerability. Her book *Gender and Wildfire: Landscapes of Uncertainty* (2013) follows people's stories of surviving, fighting, living and working with bushfire in southeast Australia and the

US west coast. She was selected by the International Social Science Council as a World Social Science Risk Interpretation and Action Fellow in 2013.

Stephen Fisher teaches in the Diploma of Community Development at Chisholm Institute, Dandenong, Australia. He has recently completed his PhD at Deakin University investigating the most effective ways to train men to become advocates for the elimination of violence against women. He has worked extensively within the Pacific and authored a Handbook on training men for women's rights, published by the Fiji Women's Crisis Centre. Stephen is also a member of the National White Ribbon Foundation Research and Policy Advisory Committee.

Kylah Genade is a Senior Lecturer in the School for Disaster Management at Stenden University in Eastern Cape, South Africa. She has worked extensively in civil society, the public sector and academia in the Caribbean and the Southern Africa region. Her focus is on vulnerability and resilience with particular emphasis on gender and children in disasters. In 2007, she developed the Girls in Risk Reduction Leadership (GIRRL) Project concept, which gained recognition as a good practice by the United Nations International Strategy for Disaster Reduction Secretariat (UNISDR) for addressing gender, climate change and disaster risk. She has been involved in developing knowledge products on topics including disaster risk assessments, Indigenous knowledge, gender and disasters, and rehabilitation and reconstruction through the African Centre for Disaster Studies and USAID. She was awarded one of two Mary Fran Myers Scholarships in 2015 in recognition of her contributions to vulnerability reduction.

Andrew Gorman-Murray is a Senior Lecturer in Social Sciences (Geography and Urban Studies) at the University of Western Sydney. His expertise is in gender geographies and in geographies of sexualities. His primary research interests are sexual minorities' experiences of belonging and exclusion in every day spaces, including homes, neighborhoods, suburbs and country towns. His work analyzes the intersections between queer politics, everyday experience and urban and regional geographies, seeking to enhance social inclusion alongside scholarly thinking. His work is published in journals across geography and the social sciences, including *Environment and Planning A*; *Social and Cultural Geography*; *Gender Place and Culture*; *International Journal of Urban and Regional Research*; *Journal of Rural Studies*; and *Antipode*. He co-edited *Material Geographies of Household Sustainability* (with Ruth Lane; 2011), *Sexuality, Rurality and Geography* (with Barbara Pini and Lia Bryant; 2013) and *Masculinities and Place* (with Peter Hopkins; 2014).

Cheryl Heykoop is a Research Associate with the Youth Creating Disaster Recovery (YCDR) Project. She recently completed a doctorate in social sciences at Royal Roads University, Canada. Her dissertation research explored how children and youth in Uganda can safely and meaningfully share their experiences of war and conflict. Cheryl is a child rights and protection consultant and advisor with the International Institute for Child Rights and

Development (IICRD), Department for Global Studies, University of Victoria. To help support and strengthen the rights and well-being of children, Cheryl has worked in various capacities with an array of organizations, including UNICEF Innocenti Research Centre, Save the Children, Right to Play, the Liu Institute for Global Issues, Plan International and various governments and academic institutions. Cheryl's research interests include child participation, children affected by conflict, transitional justice and exploring the role of local governments and communities to support and protect children.

Rachel E. Luft is Associate Professor of Sociology at Seattle University. Her primary areas of research specialization are race, gender, intersectionality, social movements and disaster. For years following Hurricane Katrina, which struck while she was teaching at the University of New Orleans, she was a participant observer in grassroots movement responses to the disaster. Much of her scholarship has focused on the racial and gender politics of New Orleans-based movement mobilization and can be found in *American Quarterly*, *The National Women's Studies Association Journal, Ethnic and Racial Studies* and other publications. She is currently writing *Disaster Patriarchy: The Intersectional Politics of the Movement for a Just Reconstruction After Hurricane Katrina.*

Kathy Lynn is a practitioner and an applied researcher with experience in working with rural, resource-based communities and Native American tribes in the Pacific Northwest to address social, environmental and economic issues associated with climate change. Kathy is part of the University of Oregon's Environmental Studies Program and coordinates the Pacific Northwest Tribal Climate Change Project. The project focuses on building an understanding of how climate change may impact the culture and sovereignty of Indigenous communities in the United States. Kathy was lead author of the chapter "Climate Change Impacts in the United States: Indigenous Peoples, Lands and Resources" within the third National Climate Assessment (2014).

Dmitriy Maslenitsyn is a sociology MA student and research assistant for the Center for Disaster and Risk Analysis at Colorado State University. He earned his BA in sociology from Pacific University in 2013. Dmitriy is currently working on his MA thesis, which examines the interplay between computer hacking, social activism and information networks. Before working on the Youth Creating Disaster Recovery Project, Dmitriy's work focused on gender issues in the context of mental healthcare professionals working with transgender clients undergoing gender transition.

Scott McKinnon is a Postdoctoral Research Fellow at the Urban Research Centre, University of Western Sydney and an Honorary Research Fellow at the University of Sydney. His postdoctoral research examines the impacts of natural disasters on sexual and gender minority populations in Australia. He has previously worked as a Research Officer in the Department of Modern History, Politics and International Relations at Macquarie University and in the School of Social Sciences and Psychology, University of Western Sydney. He has a particular interest in interdisciplinary approaches to histories and geographies of sexuality and has published in a diverse range of fields including cinema studies, memory studies, gay history, disasters research and cultural geography.

Ulf Mellström is a social anthropologist and Professor of Gender Studies at Karlstad University, Sweden. Mellström has previously held professorships in gender and technology studies and critical studies of men and masculinities. He has been an appointed research fellow at Clayman Institute of Gender Studies at Stanford University, the Department of Gender Studies at Duisberg University, Germany and the School of Social Sciences, University of Penang in Malaysia. Dr. Mellström has explored the professional culture of various groups of technicians, including civil engineers and motorcycle mechanics, based on extended periods of ethnographic fieldwork in Sweden and Malaysia. He has also developed approaches to the understanding of the gendering of technology, in particular with regard to technology and masculinity. He is author of several monographs, edited collections and articles that have appeared in leading journals. Since 2006 he has been the editor of *NORMA: International Journal of Masculinity Studies*.

Rika Morioka is a sociologist working in the field of international public health. She received her PhD from the University of California, San Diego in medical and cultural sociology. She was engaged in UN disaster responses for Myanmar's Cyclone Nargis in 2008 and Japan's triple disasters in 2011. She has lived and worked in various societies, and her multicultural perspectives orient her studies. Her research interests include social determinants of health and illness, dominant and countercultures in social change, and gender, power and hegemonic masculinity. Her publications range across a variety of topics such as recovery processes in drug addiction treatment programs in the US, death from overworking in East Asia, health activism among Japanese housewives and gender difference in perceptions toward radiation risks in post-disaster Japan. In addition to academic research, she has been active serving international nongovernmental organizations including the United Nations. She currently leads a research agency in Myanmar.

Debra Parkinson is Adjunct Research Fellow with Monash Injury Research Institute at Monash University, Australia and manager of research, advocacy and policy for Women's Health in the North and Women's Health Goulburn North East. Over the past two decades, she has researched intimate partner violence and rape, women's unequal access to the legal system and gendered discrimination through the superannuation system. Since 2009, her research has focused on environmental justice and gender and disaster. In 2015, Debra was awarded the Social and Political Sciences Graduate Research Thesis Award from Monash University for her PhD on increased domestic violence after the "Black Saturday" bushfires in Victoria, Australia.

Bob Pease is Professor of Social Work at the University of Tasmania, Australia. He has been involved in profeminist politics with men for many years and was a founding member of Men Against Sexual Assault in Melbourne.

He has published extensively on masculinity politics and critical social work practice, including four books as single author and ten books as co-editor, as well as numerous book chapters and journal articles. His most recent books include: *Undoing Privilege: Unearned Advantage in a Divided World* (2010), *Men and Masculinities Around the World: Transforming Men's Practices* (co-editor with Elisabetta Ruspini, Jeff Hearn and Keith Pringle; 2011), *Men, Masculinities and Methodologies* (co-editor with Barbara Pini; 2013) and *The Politics of Recognition and Social Justice: Transforming Subjectivities and New Forms of Resistance* (co-editor, with Maria Pallotta-Chiarolli; 2014).

Lori Peek is Associate Professor of Sociology and Co-Director of the Center for Disaster and Risk Analysis at Colorado State University. She is currently involved in a participatory project on children's post-disaster recovery in the US and Canada; a five-year project on the potential health effects of the BP Oil Spill on children; a study of long-term recovery in New Jersey after Superstorm Sandy; a study of risk perception and evacuation behavior in hurricane-prone communities along the US Gulf and Atlantic Coasts; a global examination of earthquake risk reduction activities; and a state-wide survey of disaster preparedness among childcare providers in Colorado. She is also the co-director of the SHOREline youth empowerment program and the co-leader of the Youth Creating Disaster Recovery Project. Dr Peek is author of *Behind the Backlash: Muslim Americans after 9/11* (2011), co-author of *Children of Katrina* (with Alice Fothergill; 2015) and co-editor of *Displaced: Life in the Katrina Diaspora* (with Lynn Weber; 2012).

Kylie Pybus is a graduate student at the Colorado School of Public Health with a concentration in global health and health disparities. Kylie was a graduate research assistant at the Center for Disaster and Risk Analysis from 2014 to 2015, assisting with communication and research efforts for the youth empowerment program SHOREline and the Youth Creating Disaster Recovery project.

Mark Sherry is a Professor of Sociology at the University of Toledo. His books include *If I Only Had A Brain: Deconstructing Brain Injury* (2006) and *Disability Hate Crimes: Does Anyone Really Hate Disabled People?* (2010). He is the series editor for the Ashgate Interdisciplinary Disability Studies series and his next book is *A Sociology of Impairment*. He is Chair of the Disability Section of the American Sociological Association.

Jennifer Tobin-Gurley is the Director of Research and Engagement at the Center for Disaster and Risk Analysis and a PhD candidate in sociology at Colorado State University. She earned her BA in sociology and women's studies from CSU in 2005 and MA in sociology in 2008. Jennifer's Master's research drew on qualitative interviews with local disaster recovery workers and single mothers displaced to Colorado after Hurricane Katrina. Jennifer is the recipient of the 2014 Beth B. Hess Memorial Scholarship and was chosen by CSU's

School of Global and Environmental Sustainability as a 2014–15 Sustainability Leadership Fellow. Jennifer's research interests include disaster sociology, children and youth, gender studies and qualitative research methods.

Kirsten Vinyeta is a doctoral student in the Environmental Studies Program at the University of Oregon focusing on environmental sociology. She holds a Master of Science in environmental studies from the University of Oregon. From 2011 to 2014, she was a researcher for the University of Oregon's Pacific Northwest Tribal Climate Change Project. Her research to date has primarily focused on climate change impacts and adaptation strategies within American Indian, Alaska Native and Native Hawaiian communities.

Gordon Waitt is Professor of Human Geography at the University of Wollongong, Australia. He has contributed feminist perspectives to human geography. He encourages scholars to engage with ideas that investigate the spatial imperatives of the sensuous body to explore how people make sense of self, others, things and places. He recently co-authored *Tourism and Australian Beach Cultures* (with Christine Metusela; 2012), and *Household Sustainability, Challenges and Dilemmas in Everyday Life* (with Chris Gibson, Carol Farbotko, Nick Gill and Lesley Head; 2014).

Kyle Powys Whyte is Associate Professor of Philosophy and holds the Timnick Chair in the Humanities in at Michigan State University. He is a faculty member of the Environmental Philosophy & Ethics graduate concentration and serves as a faculty affiliate of the American Indian Studies and Environmental Science & Policy programs. His primary research addresses moral and political issues concerning climate policy and Indigenous peoples and the ethics of cooperative relationships between Indigenous peoples and climate science organizations. He is an enrolled member of the Citizen Potawatomi Nation. Kyle has received the K. Patricia Cross Future Leaders Award from the Association of American Colleges and Universities (2009) and the Bunyan Bryant Award for Academic Excellence from Detroiters Working for Environmental Justice (2015).

Claire Zara was a researcher with Women's Health Goulburn North East from 2007 until 2015 when she passed away after a short illness. She was driven by her sense of justice and fairness with a goal of system change and societal change, as reflected in the research topics she chose. Her research into domestic violence after the 2009 Victorian bushfires and the effect of that disaster on men contributed significantly to policy and practice change in emergency management. Claire was pivotal in strategic change to improve disaster response to women and men, including the establishment of disaster and gender training packages for the emergency sector.

Foreword

Kenneth Hewitt

DEPARTMENT OF GEOGRAPHY AND ENVIRONMENTAL STUDIES
WILFRID LAURIER UNIVERSITY, CANADA

This book helps make visible the place and role of masculinities: it explores their social construction, how they enter all phases of the disaster cycle, and some decisive links to pre-disaster conditions. The coverage is wide geographically and in social contexts, the implications equally so. Disaster studies increasingly show how preexisting societal conditions decide exposure, vulnerability, (absent) protections and recovery chances. In turn, they depend on socioeconomic and political orders more than geophysical extremes, and are responsible for most casualties and large parts of the damage. For that reason, at least, disasters seem largely preventable (Hewitt, forthcoming) though they tend to be masked as "Acts of God," "Mother Nature's wrath" or simply "natural" hazards. The "social construction of disaster" refers to the part society plays in causes and outcomes. The term has a specific conceptual meaning: how disaster notions themselves are fashioned by and reflect the social order, especially relevant but challenging in the gender context.

Making visible, giving voice

A particular value of the work reported is to recognize, engage with and make known voices rarely heard. Authors have walked the ground to explore concerns often missing or suppressed.

In general, the social economy of risk and disasters involves various kinds of invisibility. There are things actually hidden, kept secret, masked; others are entirely visible, yet rendered socially and mentally absent, obscure or taken-for-granted (Hewitt, 1998). Hard to believe how completely women were absent, or lumped with "man," their many distinct needs and concerns overlooked in research and emergency plans. For my 1930s generation, a major turning point was Enarson and Morrow's *The Gendered Terrain of Disaster: Through Women's Eyes* (1998). For some of us, at least, that was when the blinkers started to come off. To be candid, 15 years earlier I too edited a book exploring new departures in the human ecology of disasters (Hewitt, 1983). Not one author was a woman and questions of gender were almost entirely missing.

The same year *The Gendered Terrain of Disaster* came out, I contributed to another book of which the same could be said (Quarantelli, 1998); some excellent content but all by and about "man." It was pretty typical, but no less inexcusable! Worse, I had forgotten about eminent geographers like Élisée Reclus and Peter Kropotkin who, a century earlier, had supported women's liberation and, in their writings, such early feminists and rebels as Louise Michel, citing frontline testimonies from women. Their pioneering work included almost every topic now recognized as important in disaster vulnerability, and values that frame the Universal Declaration of Human Rights. Much earlier and under far more threatening censorship, these forerunners had penned their condemnations of racism, anti-Semitism, slavery, prison systems, labor exploitation, the disenfranchisement of women, class prejudice, environmental damage, colonialism, cruelty to animals and more. Of course, along with gender, all this was marginalized by a discipline in which empire, white men, and patriarchy would prevail. And still, the present volume is bracketed by Enarson and Pease's editorial observation that "men are invisible as gendered actors in most disaster studies." How strange that this applies to masculinity—what may seem the most visible of all gendered attributes.

In disasters, as elsewhere, singular problems arise from genders being taken as "natural," fixed or physically based, when so much is done to craft them in service of vested interests and of institutional values and goals. There are invariably links between the ordinary and exceptional, the before and after the disasters. Various chapters here show how, in inverting gender relations, emergencies can potentially be opportunities for lasting change and improvement. But also, as in so many global change issues, this may not suit vested interests. They get busy, sometimes violently, trying to restore, even exaggerate, pre-disaster gender relations. Returning to things that are actually hidden, "privacy" introduces other contradictory aspects of gender and visibility. In the West, the private sphere is typically seen as a comfortable safe haven, especially for women and children, though this is also the main space of gendered or "grassroots" violence. What emerges in several chapters is how disaster tends to breach such privacy and may bring the hidden into view. The result can be threatening for masculinities, and an opening for other genders.

Critical developments in the disasters field derive from such discussions, seen most prominently, in the Yokohama Strategy, the Hyogo Framework for Action 2005–2015 and their successor, the 2015 Sendai Framework for Disaster Risk Reduction. This is reflected in an ever-growing emphasis on issues such as gender or disability, on social pre-conditions and on "underlying drivers," including gender injustice. Even so, at the beginning of the present decade, Sherilyn MacGregor (2010) could point to the poverty of women and gender studies on climate change, the most high profile environmental hazards field. She saw it as "a stranger silence," given that gender analysis is well established and when compared to the huge literature and massive expenditures on climate change research. That seems to be changing, but it is still extraordinary to me how far anthropogenic climate change has also been "nature-ized," absorbed by and interpreted through physical sciences discourse and directed by atmospheric measurement and models. The mind-set is a return to an agent-specific "hazards paradigm" (Gilbert, 1998) and a presumption of top-down, civil defense strategies and crisis management with (mainly) men in charge.

"Summits" and the male disaster power elite

Another matter, surely critical in the institutionalized forms of gender and masculinities though mostly absent or tiptoed around, begins with the fact that disaster now occurs against a background of increasingly large organizations. They favor "big" science and mega-technologies, and enterprises that are multinational, web-like, but with extreme concentrations of decision-making powers. Global governance looms large, and the global is ever more present in the local. Presiding over these developments are almost entirely male heads, boards, male elites set well *above* the glass ceilings. It is an open secret, another somehow invisible part of disaster management, yet easy to grasp, for instance in images of official summits of finance or defense ministers, CEOs, group photos of the Federal Reserve, Joint Chiefs of Staff, most Boards of Governors and others who sit at the round tables. If not all male, men are rarely less than 90% of these actors, masculinities cloaked in dark suits, uniforms or the various robes of office. I understand why practitioners working with those "at the sharp end" may lose interest in such men but disaster management depends upon these hierarchical and male-dominated institutions.

There is a strange quiet about these matters in the era of Homeland Security, the largest investment in disaster management today, where disaster preparedness is subordinate to national security and the security industrial complex (Hewitt, forthcoming). Elite male control starts with how and by whom disaster is "declared," relief funds and goods are released, and criteria are defined and applied. For disaster risk reduction, the problem is further compounded by what is done, and not done, between disasters. It is hardly accidental that, to date, the policies and priorities of these organizations have ensured that attention and funding will *not* go to the concerns of this volume (Hilhorst, 2013; Kellett and Caravani, 2013). The compelling concept of "hegemonic masculinity" seems to start in these arenas, be promoted there and go largely unchallenged. Men in command, though far removed from disaster, are integral to preparedness and whether things will improve or deteriorate, at least for others.

One of the few concerted efforts to draw this into sociological analysis was C. Wright Mills' *The Power Elite* (1956). It has no parallel that I know of in disaster literature, not even the resurgent and welcome arena of disaster politics, which yet stays carefully impersonal and institutional (Hannigan, 2010). But it is dangerous to trivialize or ignore the real male power elite behind disasters. In this case, comedy and cartoons have opened elite men to public scrutiny – not least in things that are not laughing matters and running from Aristophanes' *Lysistrata* to *Yes, Minister* and *Roger and Me*. To get any traction in the disaster world, critical analysis and journalistic exposé may also need to yield to, or start with Monty Python, *Private Eye* magazine, Comedy Central's *The Daily Show* and the like.

Social and environmental justice

Disaster studies do tend to focus on destructive events that overwhelm local communities and call in official, professional and charitable help. Research is typically directed to informing and assisting the latter, as perhaps it should be. But, in my view, the core of disaster is subjective and personal; not institutions but persons feel pain and loss. For this reason, disaster work needs a constant reminder that it is also, always, about ethics. Failed preparedness comes mainly from a cynical ignoring of those at risk, the needs and priorities of survivors and communities in distress—let alone cultural biases. It is well established that assessments and assistance should be "carried out on the basis of need alone, giving priority to the most urgent cases of distress and making no distinctions on the basis of national-ity, race, gender, religious belief, class or political opinions" (UNOCHA, 2011). The reminder may seem unnecessary except that this book is filled with discov-ered absences of such principles, sometimes but not necessarily in the conduct of humanitarian assistance.

Looking for justice does not stop with the conduct of "experts" and agen-cies. Risk reduction has to support education in, and everyday insistence upon, ethics including, when possible, dialogue shared across genders. It is hard to see this happening without dominant institutions making a larger commitment to cosmopolitan, nonviolent and conciliatory norms, rather than to bombing; to dialogue and cooperation with people living in alternative ethnic and lifestyle communities; and to gender justice as well as more inclusive gender practice.

References

Enarson, E. and Morrow, B.H. (eds.) (1998) *The gendered terrain of disaster: Through women's eyes*. Westport, CT: Greenwood.

Gilbert, C. (1998) Studying disasters: Changes in the main conceptual tools. In: Quarantelli, E.L. (ed.) *What is a disaster? Perspectives on the question*. London: Routledge, pp. 11–18.

Hannigan, J. (2010) *Disasters without borders: The international politics of natural disas-ters*. Cambridge: Polity.

Hewitt, K (ed.) (1983) *Interpretations of calamity from the viewpoint of human ecology*. London: Allen & Unwin.

Hewitt, K. (1998) Excluded perspectives in the social construction of disaster. In: Quarantelli, E.L. (ed.) *What is a disaster?* London: Routledge, pp. 75–91.

Hewitt, K. (2013) Disasters in "development" contexts: Contradictions and options for a preventive approach. *Jàmbá: Journal of Disaster Risk Studies*, 5(2), pp. 1–9.

Hewitt, K. (forthcoming) Disaster risk reduction (DRR) in the era of "homeland security": The struggle for preventive, non-violent, and transformative approaches. In: Sudmeier-Rieux, K., Jaboyedoff, M., Fernandez, M., Penna, I. and Gaillard, J.C. (eds.) *Emerging issues in disaster risk reduction, migration, climate change and sustainable develop-ment*. Heidelberg: Springer International.

Hilhorst, D. (ed.) (2013) *Disaster, conflict and society in crises: Everyday politics of crisis response*. London: Routledge.

Kellett, J. and Caravani, A. (2013) Financing disaster risk reduction: A 20 year story of international aid. Overseas Development Institute report. Available from http://www.odi.org/sites/odi.org.uk/files/odi-assets/publications-opinion-files/8574.pdf (accessed January 4, 2016).

MacGregor, S. (2010) "Gender and climate change": From impacts to discourses. *Journal of the Indian Ocean Region*, 6(2), pp. 223–238.

Mills, C.W. (1956) *The power elite*. Oxford: Oxford University Press.

Quarantelli, E. (ed.) (1998) *What is a disaster? Perspectives on the question*. London: Routledge.

UN Office for the Coordination of Humanitarian Affairs (2011) General Assembly resolution 46/182. Available from https://docs.unocha.org/sites/dms/Documents/120402_OOM-46182_eng.pdf (accessed January 4, 2016).

Part I

Critical men's studies and disaster

1 The gendered terrain of disaster

Thinking about men and masculinities

Elaine Enarson and Bob Pease

Urgent climate and disaster challenges and widening patterns of inequality and injustice at the outset of the twenty-first century call out for new ways of thinking and new partners for action. This volume offers both. Our analysis rests on an understanding of disasters as anything but "natural." Hazardous conditions and events are part of human experience, but the embedded vulnerabilities and power relations within which they occur reflect specific and discretionary development decisions. Disasters, then, are fundamentally social events that reveal social, political-economic and environmental history.[1]

Disaster risk, in turn, is best understood as a function of social and physical vulnerabilities in the face of exposure to environmental, technological, biological and human-induced or purposive hazards, both modified by people's capacity to mitigate, anticipate, adapt, resist and recover from disasters (Wisner *et al.*, 2004). Gender and sexuality matter because these social systems inexorably shape the social worlds in which disasters unfold—never unilaterally or identically, but in the complex and diverse ways illuminated in this book. It follows that how we conceptualize gender and sexuality is central to the question of disaster risk and how to reduce it. In our view, critical analysis of men and masculinities is imperative, both to flesh out the "official" gender and disaster story and as a promising platform for radical change.

We welcome readers to this volume, first with brief self-introductions and then by introducing what we mean by a critical men's studies framework for thinking about men and disasters. To further contextualize the chapters to follow, we offer a brief overview of the subfield of gender and disaster research, identify trends and gaps in past research on men and disaster, and suggest how this body of work can be used. The introduction ends with a roadmap to the organization of the book.

Personal reflections

Hurricane Andrew, the United States' most destructive disaster prior to Hurricane Katrina, struck Miami in 1992. This was my (Elaine's) first disaster experience, up close and very personal, one I experienced as an Anglo heterosexual, long-married mother of two young boys. The gender and disaster questions I've struggled with since arose from this experience, along with sharp reminders of racial and class privilege.

I was not "prepared" in any way for this event. Our family had moved from Colorado to Florida just two months earlier—long enough to put pictures on the wall but not to double-check the shutters of our new home; and long enough for my geologist husband to leave for work in rural Australia, but not to learn much from neighbors about hurricanes. With the house uninhabitable and the neighborhood unrecognizable, we temporarily relocated to north Miami where I settled the children in with new schools and doctors, stopped crying and polished up my Spanish to become a Red Cross volunteer. The long lines of exhausted women and children I saw in the early morning, waiting hour after hour for any help at all, brought home the intersectional analysis I appreciated as a feminist sociologist. I raced to the library to better understand Hurricane Andrew and what was happening to the families of Miami, only to find that the knowledge I sought about gender, race and class in disasters was about as missing in action as our insurance agent. Happily, I soon connected with new colleagues at Florida International University and began to meet with women in the eye of the storm (Morrow and Enarson, 1996; Peacock *et al.*, 1997).

Before Andrew, disaster studies were not on my radar. Long a feminist sociologist and activist, with the birth of our two sons and the opportunity to direct the University of Nevada's first women's studies program, I developed an immediate interest in masculinity. Earlier, my dissertation research had taken me to the back roads of Oregon and Washington interviewing women and men just beginning to work together in the US Forest Service in nontraditional ways, including firefighting. I also explored women's emotion work in Nevada's casinos and the underground worlds of family day care before moving to community work as coordinator of the Nevada Network Against Domestic Violence. Since 1992, these concerns have all in a different way driven my disaster research, writing and teaching, and the consulting roles I took on with women's organizations and UN agencies.

Looking back, I am struck as much by the resistance to thinking about gender and disaster as by the advances made. Can men and masculinities research break this logjam? It was a delight to engage in 2010 with a vital women's health network in Australia. Their stellar work around gender violence in bushfires led to an innovative state Gender and Disaster Task Force in which Bob Pease is strongly involved. A kindred soul at last.

Like Elaine, I (Bob) came to the field of disaster studies as a result of a personal experience of disaster, although I was not as directly impacted as Elaine was and it was much later in my academic career. In 2009 I lived in a small coastal community called Barwon Heads on the Bellarine Peninsula near Geelong in Victoria, Australia with my partner and then nine-year-old daughter. At the time I was more concerned about sea level rises in the beach-side village than other forms of disaster. The town has been historically subject to flooding and the local fish and chip shop still displayed pictures from the 1952 floods that lasted some weeks and turned the main street into a raging river.

I have been involved in critical masculinity studies for many years, arising from my experiences of profeminist men's politics in the 1970s in Australia. I have published in the fields of global and cross-cultural masculinities, men's violence against women, intersectional analyses of privilege, critical social work and profeminist practice with men. Disaster studies was not a field of scholarship that I had been involved in prior to 2009. On Saturday, February 7, 2009, a series of bushfires ignited and were burning across the state of Victoria in Australia. Referred to as the Black Saturday bushfires, 173 people died and over 400 people were injured. While I was not physically close to the fires, like many people in Victoria I knew people who were, and the loss of life caused by the fires had a ripple effect on the Victorian community.

In late 2009 I was invited to comment on a draft of a major report conducted by Women's Health Goulburn North East on the impact of the fires on women's experiences of violence. As was the case in many post-disaster contexts, men's violence against women escalated in the aftermath of the fires. Concern about the levels of increased violence by men after the bushfires led to further research interviewing the men about their experiences of the fires (see Parkinson and Zara, this volume). I was invited to be a member of the reference group on this research and later asked to bring a masculinity studies lens to men's experiences of disaster. This led to my first scholarly paper in this field (Pease, 2014). When the "Men on Black Saturday" report was launched at a Melbourne conference, Elaine and I first met and began a conversation about men, masculinities and disaster that would continue by email contact over some months. This book arose out of those conversations.

It was in the process of writing about the impact of disasters on men and how they deal with recovery that led me to think about the relevance of a gender analysis for understanding how disasters can be prevented. By exploring the contribution of critical masculinity studies to gender analysis in disaster studies, I have become more aware of the need to gender the causes of environmental crises which are often at the heart of many disasters. This led me to a broader recognition of the gendered nature of climate change and environmental destruction which I address in my single-authored chapter in this book.

The critical men's studies lens

What does it mean to bring a critical men's studies lens to disaster? We need to begin by clarifying what we mean by gender. The concept of gender was originally adopted by feminists to distinguish between the biological features of sex and the culturally constructed characteristics of masculinity and femininity (Dragiewicz, 2008). Thus gender is often discussed in the context of disaster as men's and women's socialized identities in contrast to the biological differences of sex that distinguish males from females. However, the terms "gender" and "sex" are often used interchangeably and the terms are often conflated.

Many of the references to gender in disaster studies are actually referring to sex differences, as in the case of exploring the different physical impacts that disasters have on women and men. Htun (2005) notes that most people think that gender simply refers to women and only in some cases to men. In other words, it is understood as an identity or an attribute of individuals rather than a set of practices involved in the reproduction of institutions and an attribute of social structures. Connell (2010) notes that gender is often understood simply as the categories men and women occupy in terms of sex roles or gender roles, leading to critiques she and others have developed of sex role theory (Connell, 1995; Messner, 1997; Pease, 2007). It has been noted that a major criticism of sex role theory is that it underemphasizes the economic and political power that men exercise over women and cannot explain the root causes of male domination or gender inequality (Pease, 2007).

Anderson (2005), drawing upon Risman (1998), usefully identifies three different approaches to gender: individualist, interactionist and structural. From an individualist perspective, masculinity and femininity are learned characteristics that men and women internalize into their identities. Interactionist approaches regard gender as a consequence of social interaction and enactment, whereby gender is an accomplishment (Anderson, 2005). Thus, one of the key insights of critical gender theory is that gender is something that is enacted or done rather than a state of being (Fenstermaker and West, 2002). This enactment of gender must be located in the context of larger structures and social arrangements. The focus of concern is how this doing of gender reproduces these wider structures (Pease, 2010; Schwalbe, 2014); hence, this notion captures the idea of gender not as a category but as a process whereby gender reproduces unequal gender outcomes in policies and structures. Gender is also part of the dominant symbolic order (Acker, 2006) further embedding gendered structures and practices at the interactional and institutional levels.

Structuralist approaches focus on the structural dimensions of gender through which gender is understood as a form of social structure within which persons of all genders are embedded. Htun and Weldon (2010a, pp. 5–6) define gender as

> a constellation of institutions. It is constituted by rules, norms and practices ... Gender is a feature of social structures and institutions more than human identity. It positions men and women in unequal relation of power, often intersecting (or combatting) with other institutions to uphold patterns of status hierarchy and economic inequality.

In this view, men and women are located in structurally unequal gender relations that shape and constitute their experience.

We argue that many of the studies on gender and disaster rely, in contrast, upon an individualistic approach to gender. The relational complexity of gender and its intersections with other social divisions are rarely addressed in disaster studies; but, it is critical to move beyond individualistic approaches to gender if we want to understand the complexity of the links between men, masculinities and disaster. In terms of this wider definition of gender, it may be more appropriate to think in terms of "gender order" rather than gender. Given that gender does not just include presentation of self, but also entails laws, cultural beliefs and collective practices (Schwalbe, 2014), this term captures the wider institutional dimensions of gender. Flood (2007), drawing upon Connell (1987), defines a gender order as "the patterning of gender at the level of an entire society," differentiated from gender regimes which refer to "the patterning of gender in given institutions" such as schools, workplaces or the state (p. 235). Kelly (2005) notes that the language of gender orders or gender regimes has replaced the language of patriarchy in response to criticisms of the concept. However, the concept of patriarchy is an important example of an unequal gender order where men dominate women and other men.

It is within this context that we need to locate our discussion about men and masculinities. Following Connell (2000), we argue that it is most useful to understand men and masculinities as involving these six key dimensions: 1) multiple masculinities arising from different cultures, different historical periods and different social divisions between men; 2) different positions reflected in these multiple masculinities in relation to power, with some forms of masculinity being hegemonic and dominant while other masculinities are marginalized and subordinated; 3) institutionalized masculinities embedded in organizational structures and in the wider culture, as well as being located within individual men; 4) embedded masculinities represented physically in how men engage with the world; 5) masculinities produced through the actions of individual men; and 6) fluid masculinities changing in relation to the reconstructive efforts of progressive men in response to the changes in the wider society.

Connell (2000) identifies hegemonic masculinity as the culturally dominant form of masculinity that is manifested in a range of different settings. Such masculinity is not a "thing" or a fixed identity but promoted as a desirable attainment for boys and young men to strive toward. Currently, in most parts of the world, it is presented as heterosexual, aggressive, authoritative and courageous (Connell, 2000). The manliness of men and boys is judged by their ability to measure up to this normative notion. Connell's (2000) identification of forms of marginalized and subordinate masculinities not measuring up is also useful in understanding the relationship between gender and its intersections with other dimensions of stratification such as class, race and sexuality. Connell (2000) uses these concepts to illustrate how the diversity of masculinities is marked by hierarchy and exclusion. In more recent work, she has focused attention on transnational masculinities and the global gender order (Connell, 2014a) and Southern perspectives on men and masculinities (Connell, 2014b).

Connell's concept of hegemonic masculinity is not without critique. It has been charged with being essentialist, unable to interrogate the subject, deemphasizing power and domination, naturalizing the body and reifying power (Martin, 1998; Whitehead, 1999; Holter, 2003; Hearn, 2004). However, while such criticisms raise some valid points, they fail to acknowledge the multiple usages of the concept. We believe that a renovated analysis of hegemonic masculinities proposed by Connell and Messerschmidt (2005) has ongoing usefulness as a conceptual tool for interrogating men's practices in disaster contexts. In this book we make the case for such a claim. While not all contributors adopt this perspective holistically, many acknowledge their debt to Connell's work.

Thinking about men in disasters: the context

Happily, much contemporary disaster research now accepts a notion of fully realized persons in a complicated and uncertain world, refuses the binary that locks "women" and "men" into place, and lets sexuality be as significant in disaster analysis as it is in human life. The gender and disaster scholarship and subsequent practice and policy shifts since the turn of the century as well as the much smaller body of work exploring men's lives in disaster support this.

Disaster research informed by intersectional analysis has contributed essential knowledge about family and kin relationships in crises, about social change and global development, race relations, organizational dynamics in crisis, disaster popular culture, social justice, crisis management and much more. Close analysis of the geography of crisis and essential pathways to (and against) unsustainable development brings gender analysis to the fore, including critical men's studies. In queer studies, understanding the broad social processes through which disasters are created adds to knowledge about the process of marginalization and social change movements toward inclusion and social justice. Certainly, in the fields encompassed by gender studies, we learn from disasters about how gender relations shift (or not), and about gender as a basis for privilege as well as oppression. We also learn from disaster studies that gender relations take shape in particular geographical space on a dynamic planet, and that the technological, cultural and political environments that constrain may also enable positive action in crises. In each of these areas, understanding disasters through a gender lens brings something just out of sight into view, strengthening theory and research across the disciplines in the long run; in the shorter term, we believe this has also helped to build an international community of activist-scholars with a closer eye on place, on disruption, on resistance and on progressive environmental and social change. Action-oriented researchers examining the lives of girls and women in disasters have led the way.

Gender and disaster literature in brief

The main lines of inquiry in gender and disaster research reflect the belief that gender must be systematically investigated to be meaningfully understood: Gender is never automatically, universally or necessarily the driving force in any domain of social life (for reviews, see Fothergill, 1998; Enarson and Meyreles, 2004; Enarson et al., 2007; Fordham, 2011; Enarson and David, 2012; Enarson, 2012). In disaster contexts, gender, poverty, disability, ethnicity, sexuality and other power dynamics demand comprehensive analysis, allowing for historical and comparative analysis in diverse environments and attending to interactions between agency and structure, vulnerability and capacity, empowerment and disempowerment, change and continuity. Disaster researchers and practitioners know that gender is not only in play in disaster domains presumed to be gender sensitive or in cultures presumed to be oppressive.

If early gender and disaster researchers overemphasized the vulnerabilities of women and girls, writing against a dominant trope of "no gender difference" that denied manifest inequalities disaster after disaster, much of the newer work is more nuanced, exploring women's diverse subjectivities and responses to the challenges of disaster. In part, this shift was inspired by remarkable growth in grassroots women's collaborative work as disaster risk managers, leading to more research illustrating women's leadership in mitigation, preparedness, response and long-term recovery from the effects of serious hazard events and disasters (Wickremasinghe and Ariyabandu, 2005; Dasgupta *et al.*, 2010; Enarson and Chakrabarti, 2010; Bradshaw and Fordham, 2013). Feminist work around disasters also reflects particular events in nations with strong women's networks, from India and Indonesia to Haiti, Japan and New Zealand. Over time, linkages between gender, disaster and development have also been further articulated (Bradshaw, 2013) and important advances made around genders and sexualities in disasters (e.g. McSherry *et al.*, 2014). Critical linkages between gender, disaster and climate risk are now more visible on the ground, around policy tables and in the academy (Aguilar, 2007; Alston and Whittenbury, 2013; Dunn, 2013; Enarson, 2013)—and our communities of practice are better for it. But what about the gendered terrain of disaster (Enarson and Morrow, 1998) "through the eyes of men"? The question is notably absent.

Current lines of inquiry about men and disaster

In most writing around men and masculinities, disasters are a second order concern, with little attention to place-based social crises; conversely, men are invisible as gendered actors in most disaster studies. This neglect arises, in part, from fundamental misreadings of gender, as we point out above. When the political economy of disaster risk is examined, it also arises from a misreading of capitalism as gender neutral (Acker, 2006), leading in turn to a misreading of men's lives as singularly more defined by class than are women's lives.

Though limited to date, some characteristics of men and disaster research are notable. In line with women-focused research, most studies focus on men's social vulnerabilities and on disaster impact and response. Researchers have examined why male mortality in disasters is high (Tandlich *et al.*, 2013; Jonkman and Kelman, 2005); men's crisis-driven displacement and migration (Hunter and David, 2011; Peek and Fothergill, 2008; and see Buscher, 2005); and psychosocial effects of disasters on male survivors and first responders, with attention to preexisting trauma, suicide and substance abuse (Leck *et al.*, 2007; Nishio *et al.*, 2009; Ditlevsen and Elklit, 2010; Norris *et al.*, 2002; Stuber *et al.*, 2006). Some critical insights from and about men can be found in gender mainstreaming disaster studies (Mishra, 2009); though, few studies highlight men's perspectives on such gender responsive strategies as women-first food distribution policies (but see Crawford, 2013, for the case of Haiti).

Gendered risk perception is fertile ground for researchers; for example, in studies of how men perceive, rank and tolerate environmental hazards, including the (contested) "white male effect" (Becker, 2011; Finucane *et al.*, 2000; Keenan and Hanson, 2013; Miller, 2012; Greenberg and Schneider, 1995). Findings about men also appear in studies of disaster communication (West and Orr, 2007; Major, 1999), evacuation (Bateman and Edwards, 2002) and preparedness (Kano *et al.*, 2011; Haney, *et al.*, 2007), and other core areas of applied disaster research. How disasters disrupt men's livelihoods is revealed in many development and disaster studies (e.g. Bradshaw, 2001; Paolisso *et al.*, 2002). Monographs from specific US disasters offer a close look at effects on different men's livelihoods and subjectivities, among them low-income men in Texas after Hurricanes Katrina and Rita (Reid, 2011) and fishermen hit hard by the Alaskan Exxon Valdez spill (Ritchie, 2004). Kai Erikson's early discussion (1976) of hegemonic identities and cultural norms among Appalachian men after a major dam break still sets the bar high.

Though rare, couple studies give voice to men around family and household issues in disasters (e.g. Alway *et al.*, 1998; Enarson and Scanlon, 1999). Intimate relationships are further explored, if indirectly, in studies of men and disaster-related gender violence (Sety *et al.*, 2014; and see Meshack *et al.*, 2012, on abuse committed by disaster-affected youth in Texas). More work on human trafficking is emerging from those who examine the nexus of gender, conflict and disaster, though rarely with a close focus on men (Jones, 2010). Significant work is also underway to document how people of diverse sexualities and genders are marginalized in disasters and their capacities and strengths throughout the disaster period (Stukes, 2014; Dominey-Howes *et al.*, 2014).

Significantly, gender, development and climate researchers increasingly take up core concerns from disaster studies with attention to shifts in gendered divisions of labor, gender power and violence—especially interpersonal violence, although structural violence is significant. Case studies on climate change are important guideposts for future disaster research that foregrounds men and men's relationships (e.g. Dankleman, 2011; and see Babugura, 2010, from South Africa; and Van Voorst, 2009, writing from Greenland). Hazard-specific threats such as drought are occasionally examined through a male gendered lens (Alston and Kent, 2008) as are specific high risk populations, including men marginalized by race, class and age (Klinenberg, 2002); male youth (Zahran *et al.*, 2008); widowers (Hyndman, 2008; Doppler, 2009); and first responders (Benedek *et al.*, 2007). Masculinities in firefighting are an abiding concern. Pacholok's ethnography of diverse masculine subjectivities and practices in a Canadian wildfire (2013) exemplifies this as do several contributors to this volume (and see Tyler and Fairbrother, 2013; and Thurnell-Read and Parker, 2008).

In a different vein, discourse analysis reveals the gap between facts on the ground in disasters and "myths of male superiority" in narrative accounts such as provided by Scanlon (1999) for a Canadian disaster; in studies of disaster images in popular culture (e.g. Preston, 2010; Vevea, Littlefield *et al.*, 2011; Ali, 2014); and in the representation of men and masculinities theoretically (Steinert, 2003; Melin, 2014) and empirically (e.g. in Titanic research such as conducted by Larabee, 1990).

Missed opportunities

As welcome as these new lines of enquiry are, researchers frequently fail to focus directly on men and boys or to engage contemporary masculinity theories. Potential linkages between and across disciplines are not exploited; for example, empirical research on hyper-masculine work environments (Somerville and Abrahamsson, 2003; Hall *et al.*, 2007; Ely and Meyerson, 2010; and see Martin, 2001). Analysis framed around men's "special needs" in disasters, like mainstreaming studies that account only for men's positive contributions, are inadequate. Further, the strong focus on impacts and response at the individual level gives much current men and disaster research a decidedly individualistic cast. Still lacking is structural analysis of powerful gender hierarchies and subjective gender norms as driving forces in the construction of disaster risk.

Critical masculinity studies, as understood here, can move this emerging sub-field of disaster studies forward, just as the new-to-masculinities disaster lens offers masculinity research a more place-based understanding of sex, gender and sexualities at critical life junctures. New work on disaster topics can demonstrate how male bodies and masculine subjectivities are impacted in environmental crises, and how men differently interpret, respond to and engage in disaster response and reconstruction, considering these questions at the individual, interactional, organizational, institutional and symbolic levels and with respect to disasters triggered by different hazards in diverse social and environmental contexts.

A richer, more dynamic and vastly more human portrait than currently exists of men in disasters is possible. To get there, researchers must not frame issues to inadvertently replace the "vulnerable women and powerful men" trope with its inverse. Future researchers may usefully explore how gender regimes also underpin disaster management, the relief industry, the militarized security state, international disaster and climate discourse —as well as how gender organizes the immediate experiences of men and women on the ground in disasters. Some will focus like a laser on what sustains and challenges male power in crises, and others on practical issues potentially leading to more equitable disaster policy and practice. Ideally, future work on gender and disaster concerns will draw more directly on feminist theory (Hyndman and de Alwis, 2003) and engage new communities of practice among women and profeminist men. Intersectionality will not be incidental or rhetorical in this new work but integral, as this is the essential foundation for emancipatory knowledge. Research questions must always drive method; but, context-free surveys or laboratory research on disaster topics such as leadership or decision-making may prove less useful than participatory action research (Enarson, in this volume). Cutting-edge work is sure to follow from researchers collaborating with LGBTQI and disability communities, homeless men and boys on the street, single fathers, grandfathers and men living alone, male migrants and refugees, men in violent relationships, those most at risk of economic and state violence, those threatened by environmental degradation—and all those responding creatively to these compelling circumstances.

Bringing men in and using new knowledge

The essential question for some readers of this volume may be "what next?" as applied knowledge is an asset in disaster studies. Practitioners want to know how communities at risk can become safer, and why most people fail to act, together or alone, to reduce individual or collective harm. What distinguishes different men's and women's values, practices, needs, desires and capacities in disasters and what indicators can usefully be employed to track this? How can the gender violence that may attend disasters be minimized and sanctioned? Good risk maps and sophisticated decision-making models cannot rest on partial knowledge about boys and men, women and girls. In shelter management or housing and recovery planning, neither men's nor women's particular and diverse experiences, ideas and values should be overlooked.

Practitioners taking a "whole community" approach also need to know how men self-organize in disaster contexts, including around work, political and faith values, parenting status, sexuality, military service, sports, culture and all the other "ties that bind." Disaster popular culture is potentially a powerful force in risk reduction, yet a voice still highly gendered male and hence exclusionary. Statistical and normative male dominance is difficult to miss in this field, particularly at national and international policy levels, so we need to know more about what is lost when subordinated men as well as women are marginalized. Autonomy, self-sufficiency, altruism, physical strength, bravery and stoicism are enduring themes of manliness strongly on display in disasters, but what more complicated personal stories do men have to tell?

Profeminist male researchers are important in this field—not for intellectual reasons, for the benefits of "outsider" research are well established, but for political reasons. If gender oppression is a root cause of social vulnerability, the work of empowering women and promoting gender equality is an essential step toward reducing the risk of disasters in future. Like women, legions of men around the globe are, in fact, making change—in the meanings of manhood itself and what this implies for human liberation (Pease, 2010, 2014), in hazardous neighborhoods, key social institutions, traditional disaster occupations and community projects around the globe to reduce disaster risk. With shared leadership and knowledge based on men's and women's realities equally, our collective work to advance human security, sustainable development and social justice may bear fruit, reducing human suffering in disasters and enabling new ways of living together on our volatile planet. All genders, all persons, all communities make change possible and more positive futures a reality.

Roadmap to the collection

Readers will find a rich mix of empirical work in this collection, beginning with theoretically informed contributions illuminating intersections between gender, masculinities and disaster theory and research. Bob Pease offers new ways of thinking about "man-made" disasters and men's profeminist responses, creating a strong theoretical frame for subsequent chapters and demonstrating how critical men's studies helps us understand disasters. Building on her prior work on race, class and gender, Rachel Luft critically examines intersectional gender politics, particularly the mechanisms maintaining male dominance in racially based post-Katrina movement organizing in New Orleans. Duke Austin, in turn, explores gender violence as it relates to the erosion of institutional support for hegemonic forms of masculinity. Writing about men negotiating masculinities after Hurricane

Mitch, Sarah Bradshaw brings men's experiences to life, complementing her earlier analysis based primarily on women's perspectives.

The next parts explore what men lose in disasters and with what significance. Two Australian contributors writing about bushfire open the discussion. Christine Eriksen and Gordon Waitt bring a performative lens to their study of place-based firefighting masculinity, examining how gender is created and challenged in the activities of fire management. Following on their study of women and gender violence in the wake of highly destructive 2009 bushfires, Debra Parkinson and Claire Zara document the emotional costs to men fighting these fires and to those closest to them. Two contributors extend this section on "the high cost of disaster" for men. Through three personal narratives of men widowed when the 2004 Indian Ocean tsunami hit war-stricken Sri Lanka, Malathi de Alwis challenges received wisdom about men's post-disaster substance abuse and violence, engaging readers in alternative explanations informed by psychoanalytic theory as well as her anthropological fieldwork. In turn, Rika Morioka takes us to Fukushima where she examines the historical, cultural and political-economic context for the conflict between Japanese fathers and mothers over evacuation and other choices bearing on children's health.

Contributors to Part III hone in on diversity, beginning with Mark Sherry's discussion of disability issues before, during and after disaster and how these relate to men and masculinity in disability communities and beyond. Writing about the LGBTQI community during a major flood in one of Australia's largest coastal cities, Andrew Gorman-Murray, Scott McKinnon and Dale Dominey-Howes illuminate larger patterns through close analysis of individual cases. Examining structural and cultural patterns in the US, Kirsten Vinyeta, Kyle Powys Whyte and Kathy Lynn explore factors supporting and undermining Indigenous men's vulnerability and resilience in the face of climate change, highlighting intersections between natural disasters and the political disaster of colonialism. In the final chapter of this section, Jennifer Tobin-Gurley, Robin Cox, Lori Peek, Kylie Pybus, Dmitriy Maslenitsyn and Cheryl Heykoop draw on a larger study of youth recovery to ask key questions about male youth in the US and Canada, emphasizing new lines of research around gender and age.

Finally, because we join our contributors in seeking progressive action to address climate and disaster risks, the book ends with six perspectives on making change. Fires and firefighting figure large in this collection, as in the gender, masculinity and disaster canon generally. Mathias Ericson and Ulf Mellström focus on experiences from Sweden. Their analysis complements and extends earlier chapters on Australian firefighting, analyzing how embodied relations between technologies of firefighting and masculinity exclude women and men of the "wrong kind" alike. Dave Baigent then interrogates gender mainstreaming in UK fire services, bringing both his doctoral research and personal experience over decades of fire experience to the discussion. Considering new avenues for social change in disaster and climate work, Kylah Genade reviews progressive men's organizing in Southern Africa, finding lessons and possibilities for disaster reduction; while Stephen Fisher, in turn, critiques mainstreaming training around gender and disaster management and introduces a culturally responsive approach based on his work with Pacific Islander men. Leith Dunn's chapter concludes the case studies. Writing from Jamaica about gender mainstreaming disaster risk management in the Caribbean, her chapter highlights the nexus of climate change, development, gender equality and disaster reduction as these take shape in Small Island Developing States. The section and book conclude with Elaine Enarson's action research agenda for future work on men, masculinities and disaster, organized around the need to better understand, manage and reduce disaster risk.

Finally, we are exceedingly grateful to Canadian geographer Ken Hewitt and Australian sociologist Raewyn Connell, two activist scholars who never fail to speak out for gender justice, for book-ending this volume with their reflections.

Note

1 It is a pleasure to acknowledge the many contributors who made this volume possible. Thank you for your intellectual curiosity and commitment to change. We are indebted to many at Routledge for their technical support and particularly thank Ilan Kelman, whose vision created the Hazards, Disaster Risk and Climate Change book series. Thank you, Ilan, for your leadership among men in the gender and disaster community and for your strong support of this volume.

References

Acker, J. (2006) *Class questions: Feminist answers*. Lanham, MD: Rowman & Littlefield.

Aguilar, L. (2007) *Training manual on gender and climate change*. San Jose, CR: IUCN.

Ali, Z. (2014) Visual representation of gender in flood coverage of Pakistani print media. *Weather and Climate Extremes,* 4, pp. 35–49.

Alston, M. and Kent, J. (2008) "The big dry": The link between rural masculinities and poor health outcomes for farming men. *Journal of Sociology*, 44(2), pp. 133–147.

Alston, M. and Whittenbury, K. (eds.) (2013) *Research, action and policy: Addressing the gendered impacts of climate change*. New York: Springer.

Alway, J., Belgrave, L.L. and Smith, K. (1998) Back to normal: Gender and disaster. *Symbolic Interaction*, 21(2), pp. 175–195.

Anderson, K. (2005) Theorizing gender in intimate partner violence research. *Sex Roles*, 52(11/12), pp. 853–865.

Babugura, A. (2010) Gender and climate change: South Africa case study. *Heinrich Böll Stiftung Southern Africa*. Available from http://www.boell.de/sites/default/files/assets/boell.de/images/download_de/ecology/south_africa.pdf (accessed November 2, 2015).

Balgos, B., Gaillard, J.C. and Sanz, K. (2013) The *warias* of Indonesia in disaster risk reduction: The case of the 2010 Mt Merapi eruption in Indonesia. *Gender & Development*, 20(2), pp. 337–348.

Bateman, J. and Edwards, B. (2002) Gender and evacuation: A closer look at why women are more likely to evacuate for hurricanes. *Natural Hazards Review*, 3(3), pp. 107–117.

Becker, P. (2011) Whose risks? Gender and the ranking of hazards. *Disaster Prevention and Management: An International Journal*, 20(4), pp. 423–433.

Benedek, D., Fullerton, C. and Ursano, R. (2007) First responders: Mental health consequences of natural and human-made disasters for public health and public safety workers. *Annual Review of Public Health*, 28, pp. 55–68.

Bradshaw, S. (2001) Reconstructing roles and relations: Women's participation in reconstruction in post-Mitch Nicaragua. *Gender & Development*, 9(3), pp. 79–87.

Bradshaw, S. (2013) *Gender, development and disasters*. Northampton: Edward Elgar.

Bradshaw, S. and Fordham, M. (2013) Women, girls and disasters: A review for DFID. Available from https://www.gov.uk/government/uploads/system/uploads/attachment_data/file/236656/women-girls-disasters.pdf (accessed November 2, 2015).

Burke, R.V., Goodhue, C.J., Chokshi, N.K. and Upperman, J.S. (2011) Factors associated with willingness to respond to a disaster: A study of healthcare workers in a tertiary setting. *Prehospital and Disaster Medicine*, 26(04), pp. 244–250.

Buscher, D. (2005) *Masculinities: Male roles and male involvement in the promotion of gender equality a resource packet*. New York: Women's Commission for Refugee Women and Children. Available from http://www.unicef.org/emerg/files/male_roles.pdf (accessed November 2, 2015).

Connell, R. (1987) *Gender and power*. Sydney: Allen & Unwin.

Connell, R. (1995) *Masculinities*. Sydney: Allen & Unwin.

Connell, R. (2000) The Men and the boys. Sydney: Allen & Unwin.

Connell, R. (2010) Gender, men and masculinities. Encyclopedia of life support systems, UNESCO. Available from http://www.eolss.net/Sample-Chapters/C11/E1-17-02-01.pdf (accessed November 2, 2015).

Connell, R. (2014a) Global tides: Market and gender dynamics on a world scale. *Social Currents*, 1(1), pp. 5–12.

Connell, R. (2014b) The sociology of gender in Southern perspective. *Current Sociology*, 19, pp. 1–18.

Connell, R. and Messerschmidt, J. (2005) Hegemonic masculinity: Rethinking the concept. *Gender & Society*, 19(6), pp. 829–859.

Crawford, E. (2013) 2010 Haitian earthquake: Investigation into the impact of gender stereotypes on the emergency response. MA. Oxford Brookes University

Dankleman, I. (ed.) (2011) *Women and climate change: An introduction*. London: Earthscan.

Dasgupta, S., Siriner, S. and Sarathi, P. (eds.) (2010) *Women's encounter with disaster*. London: Frontpage Publications.

Ditlevsen, D. and Elklit, A. (2010) The combined effect of gender and age on post traumatic stress disorder: Do men and women show differences in the lifespan distribution of the disorder? *Annals of General Psychiatry*, 9(32). Available from http://www.annals-general-psychiatry.com/content/9/1/32 (accessed November 2, 2015).

Dominey-Howes, D., Gorman-Murray, A. and McKinnon, S. (2014) Queering disasters: On the need to account for LGBTI experiences in natural disaster contexts. *Gender, Place & Culture*, 21(7), pp. 905–918.

Doppler, J. (2009) Gender and tsunami. PhD. University of Vienna.

Dragiewicz, M. (2008) Patriarchy reasserted: Fathers' rights and anti-VAW activism. *Feminist Criminology*, 3(2), pp. 121–144.

Dunn, L. (ed.) (2013) *Gender, climate change, and disaster risk reduction*. Working Paper # 7. Institute of Gender and Development Studies, Mona Unit, the University of the West Indies and the Friedrich Ebert Stiftung Jamaica and the Eastern Caribbean. Available from http://library.fes.de/pdf-files/bueros/fescaribe/10711.pdf (accessed November 2, 2015).

Ely, R. and Meyerson, D. (2010) An organizational approach to undoing gender: The unlikely case of offshore oil platforms. *Research in Organizational Behavior*, 30, pp. 3–34.

Enarson, E. (2012) *Women confronting natural disaster: From vulnerability to resilience*. Boulder, CO: Lynne Rienner.

Enarson, E. (2013) Two solitudes, many bridges, big tent: Women's leadership in climate and disaster risk reduction. In: Alston, M. and Whittenbury, K. (eds.) *Research, action and policy: Addressing the gendered impacts of climate change.* New York: Springer, pp. 63–74.

Enarson, E. and Chakrabarti, P.G.D. (eds.) (2010) *Women, gender, and disaster: Global issues and initiatives.* Delhi: Sage.

Enarson, E. and David, E. (2012) Introduction. In: David, E. and Enarson, E. (eds.) *The women of Katrina.* Nashville, TN: Vanderbilt University Press.

Enarson, E., Fothergill, A. and Peek, L. (2007) Gender and disaster: Foundations and directions. In: Rodriguez, H., Quarantelli, E.L. and Dynes, R. (eds.) *Handbook of disaster research.* New York: Springer, pp. 130–146.

Enarson, E. and Meyreles, L. (2004). International perspectives on gender and disaster: Differences and possibilities. *International Journal of Sociology and Social Policy,* 14(10), pp. 49–92.

Enarson, E. and Morrow, B.H. (eds.) (1998) *The gendered terrain of disaster: Through women's eyes.* Westport, CT: Greenwood Publications.

Enarson, E. and Scanlon, J. (1999) Gender patterns in flood evacuation: A case study in Canada's Red River Valley. *Applied Behavioral Science Review,* 7(2), pp. 103–124.

Erikson, K. (1976) *Everything in its path: Destruction of community in the Buffalo Creek flood.* New York: Simon & Schuster.

Fenstermaker, S. and West, C. (2002) Introduction. In: Fenstermaker, S. and West, C. (eds.) *Doing, difference, doing gender: Inequality, power and institutional change.* New York: Routledge, pp. xiii–xviii.

Finucane, M., Slovic, P., Mertz, C. K., Flynn, J. and Satterfield, T. (2000) Gender, race, and perceived risk: The "white male" effect. *Healthy Risk & Society,* 2(2), pp. 159–172.

Flood, M. (2007) Gender order. In: Flood, M., Gardner, J., Pease, B. and Pringle, K. (eds.) *International encyclopedia of men and masculinities.* London: Routledge, pp. 235–236.

Fordham, M. (2011) Gender sexuality and disaster. In: Wisner, B., Gaillard, J.C. and Kelman, I. (eds.) *The Routledge handbook of hazards and disaster risk reduction.* London: Routledge, pp. 395–406.

Fothergill, A. (1998) The neglect of gender in disaster work: An overview of the literature. In: Enarson, E. and Morrow, B.H. (eds.) *The gendered terrain of disaster.* Westport, CT: Greenwood Publications, pp. 11–25.

Goh, A. (2012) A literature review of the gender-differentiated impacts of climate change on women's and men's assets and well-being in developing countries. *International Food Policy Research Institute.* CAPRi Working Paper 106. Available from http://cdm15738.contentdm.oclc.org/utils/getfile/collection/p15738coll2/id/127247/filename/127458.pdf (accessed November 2, 2015).

Greenberg, M. and Schneider, D. (1995) Gender differences in risk perception: Effects differ in stressed vs. non-stressed environments. *Risk Analysis,* 15(4), pp. 503–511.

Hall, A., Hockey, J. and Robinson, V. (2007) Occupational cultures and the embodiment of masculinity: Hairdressing, estate agency and firefighting. *Gender, Work & Organization,* 14(6), pp. 534–551.

Haney, T., Elliott, J. and Fussell, E. (2007) Families and hurricane response: Evacuation, separation, and the emotional toll of hurricane Katrina. In: Brunsma, D., Overfelt, D. and Picou, S. (eds.) *The sociology of Katrina: Perspectives on a modern catastrophe.* Lanham, MA: Rowman & Littlefield, pp.71–90.

Hearn, J. (2004) From hegemonic masculinity to the hegemony of men. *Feminist Theory,* 5(1), pp. 49–72.

Holter, O. (2003) *Can men do it? Men and gender – the Nordic experience*. Copenhagen: Nordic Council of Ministers.

Htun M. (2005) What it means to study gender and the state. *Politics & Gender*, 1(01), pp. 157–166.

Htun, M. and Weldon, S (2010) When and why do governments promote sex equality? Violence against women, reproductive rights, and parental leave in cross-cultural perspective. Available from http://www.government.arts.cornell.edu/assets/psac/sp10/Htun_PSAC_Feb12.pdf (accessed March 29, 2016).

Hunter, L. and David, E. (2011) Displacement, climate change, and gender. In: Piguet, É., Pécoud, A. and Guchteneire, P. (eds.) *Climate change and migration*. New York: Cambridge University Press, pp. 306–330.

Hyndman, J. (2008) Feminism, conflict and disasters in post-tsunami Sri Lanka. *Gender, Technology and Development*, 12(1), pp. 101–121.

Hyndman, J. and de Alwis, M. (2003) Beyond gender: Towards a feminist analysis of humanitarianism and development in Sri Lanka. *Women's Studies Quarterly*, 31(3/4), pp. 212–226.

Jones, S.V. (2010) Invisible man: The conscious neglect of men and boys in the war on human trafficking. *The Utah Law Review*, 4, pp. 1143–1188.

Jonkman, M. and Kelman, I. (2005). An analysis of the causes and circumstances of flood disaster deaths. *Disasters*, 29(1), pp. 75–97.

Kano, M., Wood, M., Bourque, L. and Mileti, D. (2011) Terrorism preparedness and exposure reduction since 9/11: The status of public readiness in the United States. *Journal of Homeland Security and Emergency Management*, 8(1). Available from doi: 10.1016/j.hrmr.2004.10.007 (accessed November 2, 2015).

Keenan, K. and Hanson, S. (2013) Gender, place, and social contacts: Understanding awareness of vulnerability to terrorism. *Urban Geography*, 34(5), pp. 634–656.

Kelly, L. (2005) How violence is constitutive of women's inequality and the implications for equalities work. London: Equality and Diversity Forum Seminar. Available from http://www.edf.org.uk/blog/wp-content/uploads/2005/11/LizKelly2ndPaper.doc (accessed November 2, 2015).

Klinenberg, E. (2002) *Heat wave: A social autopsy of disaster in Chicago*. Chicago: University of Chicago Press.

Larabee, A. (1990) The American hero and his mechanical bride: Gender myths of the Titanic disaster. *American Studies*, 31(1), pp. 5–23.

Leck, P., Difede, J., Patt, I., Giosan, C. and Szkodny, L. (2007) Incidence of male childhood sexual abuse and psychological sequelae in disaster workers exposed to a terrorist attack. *International Journal of Emergency Mental Health*, 8(4), pp. 267–274.

Major, A.M. (1999) Gender differences in risk and communication behavior: Responses to the New Madrid earthquake prediction. *International Journal of Mass Emergencies and Disasters*, 17(3), pp. 313–338.

Martin, P. (1998) Why can't a man be more like a woman? Reflections on Connell's masculinities. *Gender & Society*, 12(4), pp. 472–474.

Martin, P. (2001) Mobilizing masculinities: Women's experiences of men at work. *Organization*, 8(4), pp. 587–618.

McSherry, A., Manalastas, E., Gaillard, J.C. and Dalisay, S.N.M. (2015) From deviant to *bakla*, strong to stronger: Mainstreaming sexual and gender minorities into disaster risk reduction in the Philippines. *Forum For Development Studies*, 42(1), pp. 27–4. Available from http://www.researchgate.net/journal/0803-9410_Forum_for_development_studies (accessed February 8, 2016).

Melin, S. (2014) Toward gendering existential risk theory. MA. Columbia University.

Meshack, A., Peters, R., Amos, C., Johnson, R., Hill, M. and Essien, J. (2012) The relationship between Hurricane Ike residency damage or destruction and intimate partner violence among African American male youth. *Texas Public Health Journal*, 64(4), pp. 30–33

Messner, M. (1997) *Politics of masculinities*. Thousand Oaks, CA: Sage.

Miller, L. (2012) Women and risk: Commercial wastewater injection wells and gendered perceptions of risk. In: Measham, T. and Lockie, S. (eds.) *Risk and social theory in environmental management*. Collingwood, VIC: CSIRO Publishing, pp. 130–146.

Mishra, P. (2009) Let's share the stage: Inclusion of men in gender risk reduction. In: Enarson, E. and Chakrabarti, P.G.D. (eds.) *Women, gender and disaster: Global issues and initiatives*. Delhi: Sage, pp. 29–40.

Morrow, B.H. and Enarson, E. (1996) Hurricane Andrew through women's eyes: Issues and recommendations. *International Journal of Mass Emergencies and Disasters*, 14(1), pp. 1–22.

Nishio, A., Akazawa, K., Shibuya, F., Abe, R., Nushida, H., Ueno, Y. and Shioiri, T. (2009) Influence on the suicide rate two years after a devastating disaster: A report from the 1995 Great Hanshin-Awaji earthquake. *Psychiatry and Clinical Neurosciences*, 63(2), pp. 247–250.

Norris, F., Friedman, M., Watson, P., Byrne, C.M., Diaz, E. and Kaniasty, K. (2002) 60,000 disaster victims speak: Parts 1 and 2. *Psychiatry*, 65(3), pp. 207–260.

Pacholok, S. (2013) *Into the fire: Disaster and the remaking of gender*. Toronto: University of Toronto Press.

Paolisso, M., Ritchie, A. and Ramirez, A. (2002) The significance of the gender division of labor in assessing disaster impacts: A case study of Hurricane Mitch and hillside farmers in Honduras. *International Journal of Mass Emergencies and Disasters*, 20(2), pp. 171–195.

Peacock, W., Morrow, B.H. and Gladwin, H. (eds.) (1997) *Hurricane Andrew: Race, gender and the sociology of disaster*. London: Routledge.

Pease, B. (2007) Sex role theory. In: Flood, M., Gardner, J., Pease, B. and Pringle, K. (eds.) *International encyclopedia of men and masculinities*. London: Routledge, pp. 554–555.

Pease, B. (2010) *Undoing privilege: Unearned advantage in a divided world*. London: Zed Books.

Pease, B. (2014) "New wine in old bottles?" A commentary on "what's in it for men?" *Men and Masculinities*. Available from doi: 1097184X14558238 (accessed November 2, 2015).

Peek, L. and Fothergill, A. (2008) Displacement, gender, and the challenges of parenting after Hurricane Katrina. *National Women's Studies Association Journal*, 20(3), pp. 69–105.

Preston, J. (2010) Prosthetic white hyper-masculinities and "disaster education." *Ethnicities*, 10(3), pp. 331–343.

Reid, M. (2011) A disaster on top of a disaster: How gender, race, and class shaped the housing experiences of displaced Hurricane Katrina survivors. PhD. University of Texas.

Risman, B. (1998) *Gender vertigo: American families in transition*. London: Yale University Press.

Ritchie, L. (2004) Voices of Cordova: Social capital in the wake of the Exxon Valdez oil spill. PhD. Mississippi State University.

Scanlon, J. (1999) Myths of male and military superiority: Fictional accounts of the 1917 Halifax explosion. *English Studies in Canada*, 24, pp. 1001–1025.

Schwalbe, M. (2014) *Manhood acts: Gender and the practices of domination*. Boulder, CO: Paradigm Publishers.

Sety, M., James, K. and Breckenridge, J. (2014) Understanding the risk of domestic violence during and post natural disasters: Literature review. In: Roeder, L. (ed.) *Issues of gender and sexual orientation in humanitarian emergencies: Risks and risk reduction*. New York: Springer, pp. 99–111.

Shows, C. and Gerstel, N. (2009) Fathering, class, and gender: A comparison of physicians and emergency medical technicians. *Gender & Society*, 23(2), pp. 161–187.

Somerville, M. and Abrahamsson, L. (2003) Trainers and learners constructing a community of practice: Masculine work cultures and learning safety in the mining industry. *Studies in the Education of Adults*, 35(1), pp. 19–34.

Steinert, H. (2003) Unspeakable September 11th: Taken for granted assumptions, selective reality construction and populist politics. *International Journal of Urban and Regional Research*, 27(3), pp. 651–665.

Stuber, J., Resnick, H. and Galea. S. (2006) Gender disparities in posttraumatic stress disorder after mass trauma. *Gender* Medicine, 3(1), pp. 54–67.

Stukes, P. (2014) A caravan of hope—gay Christian service: Exploring social vulnerability and capacity-building of lesbian, gay, bisexual, transgender and intersex identified individuals and organizational advocacy in two post-Katrina disaster environments. PhD. Texas Woman's University.

Tandlich, R., Chirenda, T.G. and Srinivas, C.S.S. (2013) Preliminary assessment of the gender aspects of disaster vulnerability and loss of human life in South Africa. *Jàmbá: Journal of Disaster Risk Studies*, 5(2), pp. 1–11. Available from http://www.jamba.org. za/index.php/jamba/article/view/84 (accessed November 2, 2015).

Thomas, D., Phillips, B., Lovekamp, W. and Fothergill, A. (eds.) (2013) *Social vulnerability to disasters*. 2nd edn. Boca Raton, FL: CRC Press.

Thurnell-Read, T. and Parker, A. (2008) Men, masculinities and firefighting: Occupational identity, shop-floor culture and organizational change. *Emotion, Space and Society*, 1(2), pp. 127–134.

Tierney, K. (2012) Critical disjunctures: Disaster research, social inequality, gender, and hurricane Katrina. In: David, E. and Enarson, E. (eds.) *The women of Katrina: How gender, race, and class matter in an American disaster*. Nashville, TN: Vanderbilt University Press, pp. 245–258.

Tolin, D. and Foa, E. (2006). Sex differences in trauma and posttraumatic stress disorder: A quantitative review of 25 years of research. *Psychological Bulletin*, 13 (6), pp. 959–992.

Tyler, M. and Fairbrother, P. (2013) Bushfires are "men's business": The importance of gender and rural hegemonic masculinity. *Journal of Rural Studies*, 30, pp. 110–119.

Van Voorst, R. (2009) "I work all the time—he just waits for the animals to come back": Social impacts of climate changes: A Greenlandic case study. *Jàmbá: Journal of Disaster Risk Studies*, 2(3), pp. 235–252. Available from http://acds.co.za/uploads/jamba/ vol2no3/van_voorst.pdf (accessed November 2, 2015).

Vevea, N., Littlefield, R., Fudge, J. and Weber, A. (2011) Portrayals of dominance: Local newspaper coverage of a natural disaster. *Visual Communication Quarterly*, 18(2), pp. 84–99.

West, D. and Orr, M. (2007) Race, gender and communications in natural disasters. *The Policy Studies Journal*, 35(4), pp. 569–586.

Whitehead, S. (1999) Hegemonic masculinity revisited. *Gender, Work and Organization*, 6, pp. 58–62.

Wickremasinghe, M. and Ariyabandu, M. (2005) *Gender dimensions in disaster management*. New Delhi: Zubaan.

Wiest, R. (1988) A comparative perspective on household, gender, and kinship in relation to disaster. In: Enarson, E. and Morrow, B.H. (eds.) *The gendered terrain of disaster*. Westport, CT: Greenwood, pp. 63–80.

Wisner, B., Cannon, T., Blaikie, P. and Davis, I. (2004) *At risk: Natural hazards, people's vulnerability and disasters*. London: Routledge.

Zahran, S., Peek, L. and Brody, S. (2008) Youth mortality by forces of nature. *Children, Youth, and Environments*, 18(1), pp. 371–388.

2 Masculinism, climate change and "man-made" disasters

Toward an environmental profeminist response

Bob Pease

There is overwhelming evidence that the world's climate is changing and that human-caused greenhouse gas emissions have made a significant impact on the climate in general and global warming in particular. Such climate change can contribute to extreme weather events such as heatwaves, floods and tropical cyclones, which in turn can create "natural" disasters. In addition to extreme weather conditions, gradual sea level rises increase flooding in coastal communities and rising temperatures will impact on water shortages and food crops.

While it may be difficult to provide scientific evidence that any one climate-related disaster such as a flood, bushfire or drought is linked with climate change, the trend over time demonstrates that there is an increasing frequency of such disasters (Boetto and McKinnon, 2013). Bouwer (2011), for example, cites research which demonstrates that while nonweather-related disasters such as earthquakes have remained constant, weather-related disasters such as floods, cyclones, storms and wildfires have increased in frequency and severity around the world in recent years. Alston (2013) also argues that many environmental disasters are caused primarily by climate change and Neumayer and Plümper (2007) maintain that nature is never the sole cause of "natural" disasters, as disasters are shaped by social, economic and cultural relations.

In light of the increasing evidence about global warming, disaster risk reduction must address the social and human causes of climate change. Helma (2006) argues that disaster studies can be usefully informed by climate change research because it locates vulnerability to disasters in the context of long-term global and local processes. Thus, taking human-induced climate change and increasing climate-related natural disasters as a backdrop, this chapter explores the gendered nature of the causes of and responses to climatic events around the world.

Gender and climate change

A growing body of literature has demonstrated that the impact of climate change is differentially felt in terms of geographical location in the world and also in terms of social locations of gender, class and race within particular geopolitical spaces (Dankelman, 2002; MacGregor, 2010a; Arora-Jonsson, 2011; Nagel, 2012). A number of commentators have also noted that much of the research on the impact of climate change on vulnerable populations, including women and those in poor communities in the Global South, has not addressed gendered power and patriarchal discourses that influence the framing of climate change as a scientific problem ostensibly unrelated to gender (Dankelman, 2002; MacGregor, 2010b; Carr and Thompson, 2014). Thus, while important research has demonstrated the differential impact on women resulting from their caring responsibilities, their economic status, their roles in food production systems and dependence on water, as well as their role as healthcare providers, there has been little consideration of how men's gender identities, especially in the Global North, shape institutional responses to climate change (MacGregor, 2010a).

Further, the gendered and intersectional dimensions of climate change have had little impact on environmental policies. There is even less research into the role of gender and other social divisions in the human causes of climate change (Bretherton, 1998; Denton, 2010; MacGregor, 2010a; Alston and Whittenbury, 2013; Kaijser and Kronsell, 2015).

While it has been noted in climate change policy forums that those in highly industrialized regions are primarily responsible for most carbon emissions, there has been little acknowledgement that men in these regions are responsible for more pollution than women (Arora-Jonsson, 2011). Two European studies in France and Finland explored the consumption patterns of women and men and demonstrated that men have a far heavier ecological footprint than women (Godoy, 2011). Men's use of transportation in private cars and use of air travel is higher than women's, as is their consumption of meat (Johnsson-Latham, 2007; Godoy, 2011).

In the context of gender differences in men's and women's ecological footprints, McCright (2010) set out to explore the differences between men and women in their understanding of climate change and their level of concern about it. He expected to find that men would have more scientific knowledge about climate change than women. However, contrary to this, he discovered that while women underestimated their knowledge about climate change, they possessed more detailed scientific knowledge than men. Numerous studies demonstrate that women report greater concern for the environment than men, higher levels of awareness of environmental problems and more interest in doing something about environmental issues, as they have historically (McCright, 2010; McCright and Sundstrom, 2013; Feygina *et al.*, 2010; Goldsmith *et al.*, 2013; Grasswick, 2014).

There are various explanations for these gender differences. Some of these explanations border on the essentialist in that they posit women as being more caring, nurturing and expressive, while men are portrayed as more competitive (Goldsmith *et al.*, 2013). However, the explanations that seem to hold most validity are those that locate the differences in men's and women's social positioning in relation to privilege and power, rather than sex differences per se. It is important to note, for example, the gender division of labor in men's greater use of cars and business flights.

Those in more privileged positions have greater motivation and more vested interest in supporting existing power relations and social priorities (Goldsmith *et al.*, 2013). This is referred to by some as "the conservative white male effect" (McCright and Dunlap, 2011; McCright and Dunlap, 2015), as conservative white men hold more economic and political power. Such men are much more prepared to accept environmental risks compared with others, especially when addressing risk may involve economic and political change that threatens their interests. Conservative white men will also have more to lose if climate change policies lead to structural changes in the economy. For these men, to address environmental problems is to challenge the existing social and political order (Feygina *et al.*, 2010). This is not to deny, of course, that there are divisions among conservative, male-dominated political parties and that upper-class white men and business interests can be contradictory. However, it is conservative white men who have also played a key role in organizing resistance to the predictions of climate science (Oreskes and Conway, 2011).

Drawing upon the feminist epistemology of situated knowledge, Grasswick (2014) also locates responses to climate change in the interests of people that arise from their social location. Supporting the research of McCright (2010), she found that white men were generally less altruistic and less likely to be concerned about the environment. However, Feygina *et al.* (2010) also place less emphasis on gender differences per se and more on membership of the most privileged groups in society where gender privilege intersects with race and class privilege and political conservatism. Some members of such privileged groups are so invested in system maintenance that they often are unable to even see the social consequences of climate change.

Somma and Tolleson-Rinehart (1997) found variations from these gender difference studies, revealing in their research that women and men who are supportive of feminism were more aware of and concerned about the environment than women and men more generally. The gender and development literature also reveals cross-cutting links with parental status and ethnicity. So gender differences between men and women in relation to climate change may be less to do with gender as a sex category and more to do with privileged positioning, critical gender consciousness and feminist awareness.

Hegemonic masculinity and climate change

Given that energy policies and programs, petrol and fossil fuel companies and funding bodies addressing climate change are all male-dominated organizations, it is important to ensure that women have access to the key decision-making bodies impacting on climate change (Dankelman, 2002; Hemmati and Rohr, 2009). However, while women have played key roles in organizing against the causes of climate change (Dankelman, 2002; Alston 2013; Boetto and McKinnon, 2013; Gorecki, 2014), there has been little interrogation of how different forms of masculinity shape men's responses to environmental concerns (Anshelm and Hultman, 2014).

Dankelman (2002) noted early on that climate change was framed as a "technical problem" that ignored the social and political dimensions of human-induced changes in the climate. Readings of the current Intergovernmental Panel on Climate Change (IPCC) reports suggest that the technical discussion still dominates these reports. However, many men with educational and professional backgrounds in engineering and the sciences who are skeptical about global warming believe that their understanding of climate change is objective and based on scientific fact (Anshelm and Hultman, 2014). Those who disagree with their particular approach to science and rationality are accused of religious fervor or "bad science." They regard their own understanding of science as being more accurate and valid than that of the scientists who raise concerns about the impact of global warming.

There is a close connection between climate change denial among some professional men and a particular form of masculinity that is grounded in engineering rationality, natural science and industrial modernization (Anshelm and Hultman, 2014). These men are thus more concerned with protecting a particular form of industrialized and patriarchal society that affirms their masculinity *and* their class privilege than they are in protecting the environment. They also believe that climate change predictions are a left-wing conspiracy to undermine society.

Hegemonic forms of masculinity promote an expectation that men enact control over themselves and other men, over women and over the environment (Schwalbe, 2009). For men to be vulnerable to external forces is to challenge their masculinity and their sense of self as a man, as being in control and taking risks are two of the key dimensions of hegemonic masculinity. Such forms of masculinity, or masculinism as Brittan (1989) characterizes it, define manhood in terms of the capacity to dominate others and the natural world. Cornall (2010) refers to the masculine mind-set as "the dominator model," which he regards as the cause of environmental problems and ecological disasters. Thus, if we want to address environmental destruction, we must address both the institutional power of culturally powerful men and also the internalization by men of the dominator model of human consciousness. Similarly, Franz-Balsen (2014) argues that hegemonic forms of masculinity are at odds with ecological sustainability. If this is so, then the promotion of environmental awareness and knowledge about the social impact of global warming will pose a threat to this form of masculinity.

Some forms of masculinity are so fragile that they have to be constantly affirmed and reproduced. If we understand these men's skepticism about climate science as a threat to their masculinity, it may open up the public debate about environmental policies and gender politics (Anshelm and Hultman, 2014). The above analysis reinforces the view articulated in the early 1990s that the current patriarchal gender regime will need to change if we are to pursue a more sustainable world (Johnsson-Latham, 2007). While gender equality is a prerequisite for sustainable development and disaster risk reduction, it can also be argued that sustainable development is a prerequisite for gender equality (Stoparic, 2006). Gender inequality is part of unsustainable development, which in turn drives climate change and increases "natural" disasters (Leduc, 2010).

There is a positive correlation between gender equality and progressive policies on climate change and environmental sustainability (Norgaard and York, 2005). This is in part explained by the greater likelihood of women's pro-environmentalism and their increased participation within state policy-making apparatuses. Furthermore, if there are common structural causes of environmental degradation and gender discrimination, greater gender equality will most likely parallel more progressive environmental policies. While this is often cited to promote improvement in the status of women, less attention is given to what it means for progressive gender and environmental politics by men.

The environmental consequences of hegemonic masculinity: responses to climate change

Just as the gender and disaster literature focuses primarily on women, MacGregor (2010b) identifies the danger of a form of feminist environmentalism that focuses on women rather than gender relations. Seeing women as "saviors," who because of their maternal role should become involved in environmental action, can reinforce the sexist representation of women; it can also act to deter men from progressive environmental action. Further, a focus on women's vulnerability in the face of climate change may not challenge men to interrogate their complicity in environmental degradation. While MacGregor (2010a) challenges feminist scholars who have neglected environmental crises, it is time for critical masculinity scholars to seriously interrogate the environmental consequences of masculinity and men's practices.

Given the wide acknowledgement among feminist scholars that hegemonic masculinity is associated with a dominator relationship to nature (Twine, 1997), it is curious that there has been so little critical masculinities scholarship on men's relationship with nature. Much of the early engagement with nature from mythopoetic writers (Bly, 1990; Moore and Gillette, 1990) had an essentialist premise that men had an essential core that was connected with nature that they needed to reclaim. Notwithstanding the essentialist writings of mythopoetic writers on "the wild man" and other mythical beings, most masculinity theorists have neglected a critical interrogation of men and the natural world.

More recently, a number of writers use the language of eco-masculinity to make the links between men and pro-environmental consciousness (Allister, 2004; Pule, 2007; Kreps, 2010). Kreps (2010) argues that eco-feminism has been unable to address men's relationship to environmentalism. To address this neglect, he proposes the term "eco-masculinities" to describe the relationship between pro-environmental behavior and masculinities. Kreps (2010) accepts the basic eco-feminist premise that most environmental crises are a product of capitalist, colonialist and masculinist policies and practices. Rejecting any essentialist notions that women are closer to nature than men, his model of eco-masculinities emphasizes the importance of gender power equality between women and men as well as changing men's dominator relationship to the natural world to one of care and coexistence. He believes that as men move toward such a new model, they will experience a greater sense of responsibility for environmental protection. While Kreps has a feminist informed view, many of the eco-masculine responses encouraging men to commune with nature seek to find redeeming features in traditional masculinity in response to eco-feminist critiques. For this reason, Twine (1997) quite rightly rejects the language of eco-masculinity.

Technological control is a key notion in new thinking about men and masculinities in the environmental context. Hegemonic masculinity involves domination, competitiveness, aggression and certainty, whereby vulnerability, emotionality and uncertainty are disavowed. For men to open themselves to fear and distress about the challenges facing the planet may hold more hope for longer-term solutions to our environmental crises. The idea that "everything is under control" (Merilainen *et al.*, 2000) can often mean that nothing is under control.

Fleming (2007) refers to the "climate engineers" who are in search of the technological fix. Such an approach is less concerned with reducing greenhouse gas emissions and more oriented toward seeking to control the weather and climate. He also ponders the potential use that governments and the military would make of such technological developments should they come to fruition. MacGregor (2010b) refers to this as the "masculinization of environmental politics," where science and global security are reflective of hegemonic masculinity. It has been noted previously that men dominate the organizations that have responsibility for addressing climate change. From the male scientists in the Intergovernmental Panel on Climate Change (IPCC) to the predominantly male celebrity spokesmen on climate change to the male leaders of the environmental lobby organizations, the politics of climate change are shaped by masculinist discourses. This is so even as female representation in science is increasing. Thus, the dominant discourse of climate change is that it is a scientific technical problem that requires technical and scientific solutions.

It does not seem contentious to argue that men are currently more oriented toward technical and scientific environmental mitigation projects (Boyd, 2010). Alaimo (2009) alerts readers to the masculinity that underpins the frameworks of scientific objectivity that purport to explain climate change science. She notes that much of global climate change science represents a transcendent view that is unable to acknowledge either vulnerability or gender and sexual diversity. Terry (2009) says that the computer modeling and technology involved in so-called solutions to climate change are stereotypically masculinist and do not allow for an understanding of the impact of gender inequality on climate change.

Buck *et al.* (2014) identify the masculinist assumptions embedded in geo-engineering projects that aim to provide technical solutions to global warming. Drawing upon representations of gender in engineering practice more broadly, they encourage feminist theorizing into engineering practices to explore the gendered limitations of the professional engineering project. These writers are revisiting the pathbreaking work of Hacker (1989) and other early feminists who theorized the relationship between gender and technology in the 1980s and 1990s.

Similar representations of a masculine mind-set are embedded in environmental management. Merilainen *et al.* (2000) identify how the very notion that the environment can be managed by technical means constitutes a rationalist ideology. Drawing upon Kerfoot and Knights' (1993) concept of the masculine mode of thinking in strategic management, Merilainen *et al.* (2000) identify how the management and control of the environment is premised on an objectifying and instrumentalist framing of environmental issues. Thus, in Western cultures, there is more discussion about the management and control of the environment (read: nature) rather than about its care and protection.

There are also similarities in the masculinity embedded in international relations and global security practice with environmental management (Ingolfsdottir, 2011). All of these fields are forms of state behavior whereby the hegemonic forms of masculinity involving power, physical strength, courage and toughness are institutionalized into state-based disciplines and policies.

Men's privilege and intersectional gender structures

Gender is easily equated with women and thus can readily ignore the wider relationship with men's domination and power. Similarly, Plumwood (1993) argues that all forms of oppression are not reduced to hegemonic masculinity. Rather, the wider dominator identity which many men occupy is also associated with other forms of privilege and dominance as well as gender.

Thus, we cannot address environmental problems without acknowledging wider social inequalities and their interrelationships (Godfrey, 2012). This view of gender is concerned with its intersection with race, ethnicity, class and other social divisions, including the relations between the geopolitical regions of North and South. Wider global inequalities impact negatively on the environment, as nations in the Global North shape the agenda and negotiations about responses to climate change (Denton, 2010) as well as producing emissions disproportionately to their population. There is thus not just a masculine bias in policies addressing greenhouse gas emissions; there is also a West-centric bias as well (Leduc, 2010).

This means that to understand people's relationship to the environment, we have to locate their responses in gender, race and class relations. Taylor (1997) notes that, notwithstanding the contribution of significant women environmentalists, many forms of environmental activism have been dominated by white middle-class men and that their concerns have shaped organized environmental responses. Because it was not inclusive of women, working-class people and people of color, their concerns were not on the environmental agenda.

While Kaijser and Kronsell (2015) argue that there has been very little engagement with the intersectionality literature from within environmental studies, this analysis has been applied to disasters (David and Enarson, 2012) and to climate change (Whyte, 2014). Nevertheless, while the gender and climate change literature is growing, very little of this new literature specifically highlights relationships between gender and related social divisions. As previously noted, gender is more often equated with women and thus women's greater vulnerability to environmental hazards becomes the focus rather than the role of men and other privileged groups in generating larger carbon footprints. There is more focus on the responses to climate change of the most disadvantaged groups than on those in the most affluent countries to environmental problems.

We thus need to shift our focus from vulnerabilities to climate change to the perpetuation of environmentally destructive norms and practices and the role in this of privileged groups in relation to class, race, gender and nationality (Kaijser and Kronsell, 2015). Further, when privileged groups engage in unsustainable practices, they represent the norm for the "good life" and become an aspirational lifestyle model for other groups. How are they able to reproduce their environmental privilege without concern for the wider problems of environmental justice?

Among others, Kari Norgaard (2012) is interested in the construction of risk in relation to environmental disasters, emphasizing people's positioning relative to class, gender, race and nationality. Norgaard argues that people's level of concern for climate change is shaped by their structural location in relation to gender, class and race. She is also interested in how people in privileged groups construct their denial about global warming. She attends to the cognitive dissonance that privileged people sustain whereby they distance themselves from the costs of their ecological footprints. While they cannot ignore the reality of climate change at one level, they manage to distance themselves emotionally from the consequences of it.

Hegemonic masculinity and environmental activism

Many environmental responses to global warming are situated within ecological modernization frameworks which unrealistically promise to reconcile ecological and industrial modern discourses in ways calling not for radical structural change but for small-scale changes. It is argued by ecomodernists that modern society does not need to be restructured to address environmental crises. Rather, economic development and modernization are falsely presented as solutions to environmental concerns (Hultman, 2013). Hultman further suggests that the ecomodernist synthesis also represents a new form of ecomodern masculinity that accommodates care for the environment within economic growth and consumption without limits. It is telling that 15 of the 18 authors of *An Ecomodernist Manifesto* are men (Asafu-Adjaye *et al.*, 2015).

Hegemonic masculinity is under threat from climate change. It remains to be seen whether men will be able to respond to this threat to their masculinity with a new subjectivity that acknowledges the global issues at stake, or whether masculinist defensive reactions will prevail. Over 25 years ago, Connell (1990) interviewed six men involved in environmental campaigns to explore the extent to which they were engaged in remaking their masculinity as a result of their environmental consciousness. While the environment movement at that time was not particularly gender conscious, and it is only marginally so now, the involvement of feminist women and the feminist collectivist processes that were integral to the politics of the movement encouraged the men to challenge traditional forms of masculinity.

Similarly, drawing upon data from an Australian Electoral Study in 2001, Leyden (2005) notes the positive correlation between environmental consciousness and support for gender equality. While only 30% of Liberal voters thought that gender equality reforms had not gone far enough, 82% of Australian Green voters believed this to be the case. It thus seems that men who demonstrate an environmental consciousness are also more likely to be progressive on gender issues. This positive correlation is supported by Stoddardt and Tindall's (2011) research in British Columbia, where they interviewed environmental activists to see what connections they made between environmental politics and feminism. Their study participants overwhelmingly cite hegemonic masculinity as the main reason why men are not more concerned about the environment. If this is so, then engaging men in reconstructing masculinity is critical to promoting men's environmental activism. Furthermore, it suggests that profeminist activists who are challenging men's violence against women should also explore the connections between the oppression of women and environmental destruction, as early feminists asserted.

Meat eating men: alternative masculinities and sustainability

Rothgerber (2013) notes research which demonstrates that eating meat and the consequence, farming of animals for consumption, contributes more to global warming than the carbon emissions of all transport. If hegemonic masculinity is constructed on the premise of controlling nature, as well as controlling women and other men, then sustainability practices challenge both hegemonic masculinity and patriarchy, if only indirectly, and help reduce environmental damage.

Meat eating is a case in point. Only a small number of people in Western cultures do not eat meat and most of those who do eat meat are men. Twine (1997) asks what alternative identities men will need to take on if they are to reconfigure their relationship with nature. Like others, he believes that environmentalism poses a significant threat to hegemonic masculinity. This is demonstrated by, among other patterns, the negative social responses male vegetarians experience, and men who more generally demonstrate ethical concern for animals and the environment. Rogers (2008) analyzes advertisements promoting meat consumption to reveal the gendered symbolism embedded in the eating of meat.

Given the great percentage of men who eat meat worldwide, Rothgerber (2013) ponders the relationship between meat consumption and masculinity. He argues that there is a resurgence in a "meat-as-masculinity discourse," whereby meat consumption is a symbol of patriarchal masculinity. Vegetarian men thus violate normative expectations of masculinity's association with meat consumption. Given the impact that meat consumption has on carbon emissions, resistance to vegetarianism among men because of the threat it might pose to hegemonic masculinity is a matter of concern. As one of the motivations for vegetarianism is the environmental impact of meat consumption, meat eating by men as a means of shoring up traditional masculinity and demonstrating virility is another gender-based obstacle to environmental sustainability.

Because green attitudes are viewed by some as "feminine," some environmentalists argue that climate change solutions should quite deliberately be presented as more masculine in order to engage greater numbers of men. This is the view proposed by Friedman (2007), who argues that the environment movement should also market itself in a more masculine mode to break the perceived connection between environmental consciousness and softness or femininity in men. However, if the environment movement is aiming to avoid challenging men's privileges, it is hard to imagine how it can address the causes of the environment crisis. We should not marginalize what are seen as "feminine" values in order to recruit more men to environmentalism, as such values are essential in addressing the causes of environmental problems.

Conclusion

How do we build a world in which men are no longer in control of women and the environment? How do we explore a way of men being in the world where they no longer have to deny their emotional and physical vulnerability? Such a project stands in sharp contrast to the mythopoetic notion of romanticizing men as warriors and hunters, as if this were a true representation of men's relationship with nature. Responding to these challenges will mean we will need to break the nexus between men's gender identity and masculinist forms of gender power.

If, as I have argued here, there is a relationship between the exploitation of nature and hegemonic masculinity, profeminist men involved in reconstructing or exiting dominant forms of masculinity may be able to envisage new nonoppressive ways for men to relate to nature, as they discover new ways to relate to women, other men and themselves. We have inspirational examples of men challenging men's violence against women; opposing pornography and men's sexual exploitation of women; creating new forms of caregiving masculinities in relation to children and other caring roles; and interrogating and relinquishing privileged positions that depend on complicity in the oppression of others. As this chapter has argued, and as feminists have long demonstrated, progressive environmental politics is not free of hegemonic masculinity and masculinism. This is why there must be a strong profeminist engagement with environmentalism. Profeminist men now urgently need to forge new alliances with environmental activists and encourage a rethinking of masculinity and men's practices in the environment movement. The ideas presented here are a clarion call for such an engagement. Arguably, the viability of the planet is at stake if women and men together do not act now to engage positively with men to address the consequences of men's destructive practices.

References

Alaimo, S. (2009) Insurgent vulnerability and the carbon footprint of gender. *Kvinder Kon Forskning NR*, 3–4, pp. 22–35.

Allister, M. (2004) Introduction. In: Altister, M. (ed.) *Eco-man: New perspectives on masculinity and nature*. Charlottesville, VA: University of Virginia Press, pp. 1–13.

Alston, M. (2013) Environmental social work: Accounting for gender in climate disasters. *Australian Social Work*, 66(2), pp. 218–233.

Alston, M. and Whittenbury, K. (eds.) (2013) *Research, action and policy: Addressing the gendered impacts of climate change*. New York: Springer.

Anshelm, J. and Hultman, M. (2014) A green fatwar? Climate change as a threat to the masculinity of industrial modernity. *NORMA: International Journal for Masculinity Studies*, 9(2), pp. 84–96.

Arora-Jonsson, S. (2011) Virtue and vulnerability: Discourses on women, gender and climate change. *Global Environmental Change*, 21, pp. 744–751.

Asafu-Adjaye, J. et al. (2015) *An ecomodernist manifesto*. Available from www.ecomodernism. org (accessed March, 14 2014).

Bly, R. (1990) *Iron John: A book about men*. New York: Routledge.

Boetto, H. and McKinnon, J. (2013) Gender and climate change in rural Australia: A review of differences. *Critical Social Work*, 14(3), pp. 15–31.

Bouwer, L. (2011) Have disaster losses increased due to anthropogenic climate change? *American Meteorological Society*, January, pp. 39–46.

Boyd, E. (2010) The Noel Kempff project in Bolivia: Gender, power, and decision-making in climate mitigation. *Gender and Development*, 10(2), pp. 7–77.

Bretherton, C. (1998) Global environmental politics: Putting gender on the agenda. *Review of International Studies*, 24, pp. 85–100.

Bretherton, C. (2003) Movements, networks, hierarchies: A gender perspective on global environmental governance. *Global Environmental Politics*, 3(2), pp. 103–119.

Brittan, A. (1989) *Masculinity and power*. London: Basil Blackwell.

Buck, H., Gammon, A. and Preston, C. (2014) Gender and geoengineering. *Hypatia*, 29(3), pp. 651–701.

Carr, E. and Thompson, M. (2014) Gender and climate change adaptation in agrarian settings: Current thinking, new directions and research frontiers. *Geography Compass*, 8(3), pp. 182–197.

Connell, R. (1990) A whole new world: Remaking masculinity in the context of the environment movement. *Gender & Society*, 4(4), pp. 452–478.

Cornall, P. (2010) Sustainable masculinity is interwoven with environmental sustainability. *Peace and Collaborative Development Network* blog, April 27.

Dankelman, I. (2002) Climate change: Learning from gender analysis and women's experience of organizing for sustainable development. *Gender and Development*, 10(2), pp. 21–29.

David, E. and Enarson, E. (eds.) (2012) *The women of Katrina: How gender, race and class matter in an American disaster*. Nashville: Vanderbilt University Press.

Denton, F. (2010) Climate change vulnerability, impacts and adaptation: Why does gender matter? *Gender and Development*, 10(2), pp. 10–20.

Feygina, I., Jost, J. and Goldsmith, R. (2010) System justification, the denial or global warming and the possibility of "system sanctioned change." *Personality and Social Psychology Bulletin*, 36(3), pp. 326–338.

Fleming, J. (2007) The climate engineers. *The Wilson Quarterly*, Spring, pp. 46–60.

Franz-Balsen, A. (2014) Gender and (un)sustainability: Can communication solve a conflict of norms? *Sustainability*, 6, pp. 1974–1991.

Friedman, T.L. (2007) The power of green. *New York Times Magazine*, April 15.

Godfrey, P. (2012) Introduction: Race, gender and class and climate change. Race, Gender and Class, 19(1–2), pp. 3–11.

Godoy, J. (2011) Climate change in Europe: Pollution is a masculine noun. *Tierramerica: Environment and Development*, February 17.

Goldsmith, R., Feygina, I. and Jost, J. (2013) The gender gap in environmental attitudes: A system justification perspective. In: Alston, M. and Whittenbury, K. (eds.) *Research, action and policy: Addressing gendered impacts of climate change*. New York: Springer, pp. 159–173.

Gorecki, J. (2014) No climate justice without gender justice: Women at the forefront of the people's climate march. *The Feminist Wire*, September.

Grasswick, H. (2014) Climate change science and responsible trust: A situated approach. *Hypatia*, 29(3), pp. 541–557.

Hacker, S. (1989) *Pleasure, power and technology*. Boston: Unwin Hyman.

Helma, M. (2006) Natural disasters and climate change. *Disasters*, 30(1), pp. 1–4.

Hemmati, M., and Rohr, U. (2009) Engendering the climate-change negotiations: Experiences, challenges and steps forward. *Gender and Development*, 17(1), pp. 19–31.

Hultman, M. (2013) The making of an environmental hero: A history of ecomodern masculinity, fuel cells and Arnold Schwarzenegger. *Environmental Humanities*, 2, pp. 79–99.

Ingolfsdottir, A. (2011) Climate change and security in the Artic: The promise of feminism. *Nordic Information on Gender*. Available from httm://www.nikk.no/en/climate-change-and-security- in-the-arctic-the-promise-of-feminism (accessed November 9, 2015).

Johnsson-Latham, G. (2007) A study on gender equality as a prerequisite for sustainable development: Report to the Environment Council, Sweden.

Kaijser, A. and Kronsell, A. (2015) Climate change through the language of intersectionality. *Environmental Politics*, 23(3), pp. 417–433.

Kerfoot, D. and Knights, D. (1993) Management, masculinity and manipulation: From paternalism to corporate strategy in financial services in Britain. *Journal of Management Studies*, 30(4), pp. 659–677.

Kreps, D. (2010) Introducing eco-masculinities: How a masculine discursive subject approach to the individual differences theory of gender and IT impacts an environmental informatics project. Americas Conference on Information Systems, Proceedings, Paper 277.

Leduc, B. (2010) Climate change and gender justice. *Climate and Development*, 2(4), pp. 390–392.

Leyden, S. (2005) Reconstruction and resistance: Masculinity, gender and relationships among men in the environment movement. PhD. Swinburn University of Technology.

MacGregor, S. (2010a) A stranger silence still: The need for feminist social research on climate change. *Sociological Review*, 57, pp. 124–140.

MacGregor, S. (2010b) Gender and climate change: From impacts to discourses. *Journal of the Indian Ocean Region*, 6(2), pp. 223–238.

McCright, A. (2010) The effects of gender on climate change knowledge and concern in the American public. *Popular Environment*, 32, pp. 66–87.

McCright, A. and Dunlap, R. (2011) Cool dudes: The denial of climate change among conservative white males in the United States. *Global Environmental Change*, 21, pp. 1163–1172.

McCright, A. and Dunlap, R. (2015) Bringing ideology in: The conservative white male effect on worry about environmental problems in the USA. *Journal of Risk Research*, 16(2), pp. 211–226.

McCright, A. and Sundstrom, A. (2013) Examining gender differences in environmental concern in the Swedish general public, 1990–2011. *International Journal of Sociology*, 43(4), pp. 63–86.

Merilainen, S., Moisander, J. and Personen, S. (2000) The masculine mindset of environmental management and green marketing. *Business Strategy and Environment*, 9(3), pp. 151–162.

Moore, R. and Gillette, D. (1990) *King warrior, magician lover: Rediscovering the archetypes of the masculine*. San Francisco: Harper.

Nagel, J. (2012) Intersecting identities and global climate change. *Identities: Global Studies in Culture and Power*, 19(4), pp. 467–476.

Neumayer, E. and Plümper, T. (2007) The gendered nature of natural disasters: The impact of catastrophic events on the gender gap in life expectancy, 1981–2002. *Annals of the Association of American Geographers*, 97(3), pp. 551–566.

Norgaard, K. (2012) Climate denial and the construction of innocence: Reproducing transnational environmental privilege in the face of climate change. *Race, Gender and Class*, 19(1–2), pp. 80–103.

Norgaard, K. and York, R. (2005) Gender equality and state environmentalism. *Gender & Society*, 19(4), pp. 506–522.

Oreskes, N. and Conway, E. (2011) *The merchants of doubt*. New York: Bloomsbury Press.

Plumwood, V. (1993) *Feminism and the mastery of nature*. London: Routledge.

Pule, P. (2007) Ecology and environmental studies. In: Flood, M., Gardner, J., Pease, B. and Pringle, K. (eds.) *International encyclopedia of men and masculinities*. London: Routledge, pp. 158–162.

Rogers, R. (2008) Beats, burgers, and hummers: Meat and the crisis of masculinity in contemporary television advertisements. *Environmental Communication*. 2(3), pp. 281–301.

Rothgerber, H. (2013) Real men don't eat (vegetable) quiche: Masculinity and the justification of meat consumption. *Psychology of Men and Masculinity*, 14(4), pp. 363–375.

Schwalbe, M. (2009) Denormalizing the signs of impending disaster. *Common Dreams*. April 18. Available from www.commondreams.org (accessed February 21, 2016).

Somma, M. and Tolleson-Rinehart, S. (1997) Tracking the elusive green women: Sex, environmentalism, and feminism in the U.S. and Europe. *Political Research Quarterly*, 50 (March), pp. 153–169.

Stoddardt, M. and Tindall, D. (2011) Ecofeminism, hegemonic masculinity and environmental movement participation in British Columbia, Canada, 1998–2007: "women always clean up the mess." *Sociological Review*, 31, pp. 342–368.

Stoparic, B. (2006) Women push for seats at climate policy table. Womensenews.org, July 6. Available from http://womensenews.org/story/environment/060706/women-push-seats-at-climate-policy-table (accessed November 9, 2015).

Taylor, D. (1997) American environmentalism: The role of race, class and gender in shaping activism 1820–1995. *Race, Gender and Class*, 5(1), pp. 16–62.

Terry, G. (2009) No climate justice without gender justice: An overview of the issues. Gender and Development, 17 (1), pp. 5–18.

Twine, R. (1997) Masculinity, nature and ecofeminism. *ecofem.org/journal*. Available from http://richardtwine.com/ecofem/masc.pdf (accessed November 9, 2015).

Whyte, K. (2014) Indigenous women, climate change impacts and collective action. *Hypatia*, 29(3), pp. 599–616.

3 Men and masculinities in the social movement for a just reconstruction after Hurricane Katrina

Rachel E. Luft

Gender and disaster scholars note that during and after disaster traditional gender roles and patterns can be either exaggerated or subverted (Enarson, 2012; Pacholok, 2013). The temporary dissolution of normal life can facilitate the reversion to extreme forms of the gender binary or, conversely, the transgression of normative arrangements and production of new opportunities for gendered practice. While the literature indicates that regression to hierarchical gendered power dynamics is more common, transformation is sometimes possible. This chapter examines both the reinstallation of, and intervention into, dominant gendered patterns after Hurricane Katrina struck the Gulf Coast on August 29, 2005. The setting is the New Orleans-based, grassroots social movement that fought for a just reconstruction. Instead of focusing on gender identity, I explore the gendered mechanisms that differentially advanced men into movement leadership, and also the gendered symbolic meanings of ostensibly gender-neutral practices. I center men and masculinity in an intersectional framework, uncovering the ways in which both were privileged and sometimes transformed in movement activity in the years after the disaster. My interest is the practices that promote the dominance of men and masculinity after crisis as well as those that work to destabilize it.[1]

Context and approach

Within hours of Katrina's landfall, local and national grassroots social movement organizers coalesced in the effort I call the Movement for a Just Reconstruction. The movement resembled other post-disaster civil society emergent groups in advancing a recovery agenda (Enarson, 2012). In significant ways, however, it also differed from them in using the language and tools of social justice as organizers sought to reframe the disaster of the hurricane as a product of long-standing racial and economic oppression. The movement politicized hurricane-related grievances and linked the disaster to ongoing US social crises. It focused on seven primary areas: grassroots relief and recovery, displacement or the right of return, the reconstruction of affordable housing and healthcare, worker and immigrant justice, and criminal justice reform.

As the site of this study, the post-disaster social movement provides the opportunity to examine men and masculinity across other social positionalities; important intersections of race, class, place, age and political experience complicate reductive notions of gender while also revealing gendered patterns. Additionally, the Movement for a Just Reconstruction brought together local disaster survivors and nonlocal activists. The former includes men who were living in New Orleans at the time of the hurricane and whose lives were personally disrupted by the disaster; the latter includes people I call activist-volunteers, supporters from around the country who came after the storm to participate in the movement.

The two primary foci of the chapter are the advancement of men to positions of movement leadership and the promulgation—and occasional subversion—of dominant modalities of masculinity in movement values, visions and organizing styles. By men I mean people who accept their early male gender assignment. By masculinity I refer to normative "pattern[s] of practice" that privilege certain values, enactments and social relations over others in a hierarchy of gender power (Connell and Messerschmidt, 2005, p. 832).

I am a White, feminist sociologist who had been involved in racial justice movements for years before the hurricane struck. I was living in New Orleans at the time of the disaster and spent the next six years conducting participant observation in post-Katrina movement groups. I interviewed or held focus groups with forty movement leaders and activists, plus seven others who worked in non-profit or related capacities. Thirty-one were women, fifteen were men, and one was genderqueer. Twenty-five were Black, nineteen White, and three Latino/a. There were twenty-nine hurricane survivors and eighteen who were nonlocal. The respondents featured here were very involved in the loose network of post-disaster movement groups. While I put men and masculinity at the center of this study, I draw heavily from the perspectives of women and the gender nonconforming people who worked with them. I begin by describing the reproduction of men's leadership, turn to an analysis of movement masculinity and then address instances of actual and potential intervention.

Men in leadership: the glass forklift

Scholars of gender and work have identified a "glass escalator" effect in the workplace that promotes (some) men into positions of leadership (Williams, 2013). In this section I suggest that the gendered conditions of disaster in combination with the gendered politics of social movement activity can facilitate men's leadership in post-disaster movement formations in a process I call the "glass forklift." Here we see the normal advantages to men in the paid workforce, public sphere and social movement activity—lack of caregiving responsibilities, default masculinist practices and so forth—exaggerated and multiplied in the disaster setting.

The devastation of infrastructure after Hurricane Katrina had differential effects for differently gendered people. Since men take less responsibility for caregiving labor, the absence of adequate housing, medical care, schools and transportation was frequently simpler to navigate than it was for women who care for children, the elderly and the sick. Men were often able to return to the city more quickly and were less consumed with managing basic needs for themselves and their families.

When they did face personal recovery tasks they were more likely to receive help from the movement in ways that would turn them into people who could organize full time. Activists organized work parties to help male leaders get up and running: they cleared and cleaned storm damage, did home repair, gutted flooded residences, provided rides and, in the case of at least one local male organizer, performed childcare. The personal recovery needs of male leaders, that is, were interpreted to be politically salient. In this way, disaster relief and selective occasions of reproductive labor were reframed as political activism and practiced in gendered ways.

Conversely, women organizers in New Orleans were rarely identified as critical movement leaders for whom personal support was political. They described coming back to the city months after male leaders were already reestablished in stable living conditions and therefore in movement leadership as well. Organizers knew they had a small window of public attention after the hurricane; those who were installed first disproportionately shaped movement organizations, set agendas and directed financial resources. For example, a Black feminist woman organizer had been heartbroken not to be able to return to the city immediately, as she had to find housing and school for her young son. Despite this obstacle, she was instrumental in getting one of the larger emergent movement groups off the ground. When she and her child finally made it back to New Orleans in January 2006, she was offered a paid staff position with the group she had helped to found. She asked for one week to find an open school or childcare for her child in the devastated city and was told the organization could not wait; they offered the position to someone else. Bill Quigley, a White movement lawyer, summarized the consequences of these practices:

> The leadership of the response ... the people who had jobs ... who got money ... [it was a] very high proportion of men who did that and yet for every group, the group [of movement rank and file] was 80–90% women.

As women juggled movement organizing with caregiving work, men were free(d) to spend long hours in strategy sessions, national conference calls and public events. They were recognized as leaders during the early days and weeks when the press was paying particular attention and activist-volunteers from around the country were seeking direction about how to help. Several of these men turned very early to the media, which gave them national recognition as movement leaders. For example, Jordan Flaherty, a White male anarchist and labor organizer, wrote a long description of his experiences in the first days after Katrina. The piece went viral in radical circles across the country and soon he was receiving press calls for interviews. He describes how, guided by feminist, antiracist principles, he first turned down the interviews, recommending instead several local women of color movement organizers. Later, to his chagrin, he learned that the media had not made contact with the women, instead interviewing several other White men who did not share his political commitments. From then on, he agreed to speak to the press and the experience transformed him into an independent journalist with an international reputation.

Jordan was not the first person to mention that it was often difficult to reach local women organizers by phone. The politics of visibility, accessibility and personal infrastructure are all deeply gendered and intersectional. Shana griffin, a local Black feminist woman, was scolded by more than one White feminist philanthropist for being too busy to meet with potential donors. It was an exasperating charge. Shana was a key movement organizer in a struggle that had enormous external challenges and myriad internal sexist dynamics, which meant she was fighting on two fronts. She was also parenting a child while being constantly on call for a steady stream of "helpers" (e.g. the press, donors, eager volunteers). It would have been impossible to respond to the endless barrage of demands. As her situation and that of others after the storm demonstrates repeatedly, it takes infrastructure to create infrastructure, and access to infrastructure is gendered and racialized, both personally and organizationally. Where activist-volunteers arrived at male leaders' homes with tools, a car or an offer of physical labor, well-intentioned people showed up to "help" Shana with a list of questions and a need for information. The end result was that some male leaders received a continuous flow of contributions—clean-up, childcare and administrative labor—allowing them to have a vocal local and national presence, which in turn increased their national profile, funds and local power.

The temporary collapse of distinctions—between public and private, between personal reproductive labor and collective movement activism—that happened disproportionately for men after the disaster has the potential to be a model for refiguring notions of movement activity and culture as well as divisions of labor in daily life. Helping people become stable so that they can lead is a powerful reformulation of political practice with potentially radical gendered implications. After Katrina, however, it was enacted in ways that served to bolster men's power and marginalize women leaders.

There were myriad other gendered practices that interacted with the post-storm environment to solidify men's organizational power. Forms of organizational and movement sexism that abound in non-disaster settings were exacerbated by the conditions of post-hurricane life. Rosana Cruz, a Latina feminist, chronicled some of them: "I definitely recognized that there were mad gender dynamics happening all around me. There was the dynamic of the macho model of organizing, there was the gross organizing-with-your-dick dynamic." She described charismatic male leaders who "just romanc[e] people, specifically women, into doing stuff [for them].... [They are] very intellectual and magnetic, right.... you make women in different situations fall in love with you so that they'll do what you want." Others also portrayed the reign of charming men who were in prominent public positions while relying on women's support labor behind the scenes. Khalil Shahyd, a working-class Black man from Louisiana, described these dynamics. One group, for example:

> emphasized certain works, basically the house gutting which was done by all the men, this physical labor … this is the chest beating labor, this is the valorous labor.... But yet you know the work that other partners were doing, whether it was Keneika, who was managing everything in the offices,

> whether it was Shana and other folks who [were] actually managing the
> finances and the resources, whether it was Althea and those folks who cre-
> ated Safe Streets ... whether it was Shana again trying to create this woman
> of color health center ... there was no communication from [the organiza-
> tion] to the outside world about any of it.... I tried to ... show them how the
> gender dynamic was very uneven in the way that the organization ... was
> functioning, but they were not willing to hear that at all.

The gendered experience of disaster interacts with normative organizational and
cultural gendered patterns to promote men's leadership during times of crisis.
Identifying preexisting gendered structures and post-disaster gendered emergent
response patterns are necessary to interrupt what otherwise appears to be the natural advancement of men over women.

Masculinization of a movement: from gendered people to gendered practices

In most cases, the installation of men as the public, nationally recognized, funded
leaders of post-Katrina emergent movement organizations promoted a highly gendered organizing vision and set of strategies and practices. Despite the masculinization of movement choices, they were overwhelmingly coded as gender neutral. In
this way, men in power are conduits for the promulgation and institutionalization of
gendered culture while cloaking this approach in gender-blind language. Even those
who recognize the numerical disparity of gendered bodies in leadership may have
difficulty perceiving the gendered dynamics of movement culture. Jordan expressed
the challenge this way:

> I always had this framework of race and gender, so of trying to lift up people
> of color voices and within that lift up women's voices.... But I think that the
> next step of ... what does it mean to really make gender a crucial part of this
> struggle.... I'm still trying to figure that out.

The move in a gender audit from centering gendered people to centering gendered practices is difficult under any circumstance. Disasterscapes, however,
are particularly challenging contexts in which to do so because they are already
gendered in the popular imagination (Luft, 2008; Enarson, 2012). Normative
ways of making sense of disaster settings, that is, are filtered through gendered
meanings. The post-Katrina disasterscape was chaotic, desolate and heavily militarized. Organizers reflected on the gendered logic with which it was produced
and narrativized, and identified the gendered consequences of these frames for
the movement. As Kai Barrow, a Black feminist woman who came soon after the
hurricane, recalled:

> There were no children, very few women, there were no grocery stores ...
> it was barren.... there was dead bodies.... it felt raped, it really did. I say

all that to say … immediately what … emerged … it was very much about shutting down, you know, shutting down and locking in a particular modality. It wasn't about opening up, it wasn't about creative thinking…. Or this idea around intersectionality: as we talk about race can we also look at where gender and sexuality and class come into play?

Kai describes how the gendered experience of the disasterscape was reified in ways that promoted male leadership and naturalized masculinity. What emerged were gendered standards for ideal typical organizers and gendered notions of what counted as activism, both of which were presented in gender-neutral ways. Rosana gave an example from the group that was organizing day laborers against immigration and employment injustice:

> There is this gender dynamic of "well, the workers, who are men, are living in these [unsafe conditions in flooded, moldy homes or tents] and so it's OK for us to mirror their living situation." Unless you have a kid or you're taking care of people or you're in this caretaking position as a woman [in which case you cannot]. So then also the meetings that would happen where decisions would be made were in a bar at 11 o'clock at night or like all of these situations where it was gendered just "by circumstance," quote unquote.

   The engendering of political activity did not simply promote male organizers, but defined the very terms of organizing: what counts as critical, what should be prioritized, how scarce resources should be utilized and distributed. Apparent martyrdom and all-or-nothing activism were gendered and they produced gendered results, creating standards by which movement effort was evaluated. People who could not meet these expectations, because of caregiving responsibilities, political differences or other reasons, were perceived to have a lack of devotion to the movement. Rosana told the story of a woman who was organizing mostly Latino day laborers. Despite the fact that the woman spoke Spanish when some of the male leadership did not, she was marginalized in the organization because she

> had two small children and so basically there wasn't really any kind of accommodation [such as] let's try and figure out how this person who can actually communicate with the workers because she speaks Spanish, how can we make this work…. [Instead] it's like [she's] not reliable, she has an issue, then suddenly she became vilified…. [But] she was [just] a mom.

Rosana's observation that the ways in which male leaders' limitations—such as not speaking Spanish—were minimized while women's gendered responsibilities were perceived to be liabilities demonstrates the gendered social construction of the movement activist. It produced and valued certain kinds of organizers and certain kinds of organizing practices.

The compelling conditions of the disasterscape and the narrow window of opportunity for redress produced a palpable mood of political urgency in the movement. While there was a material dimension to this reality, urgency was constructed in gendered ways with gendered effects. Rosana described this process:

> I think especially after Katrina, it felt like people did not have time for a gender analysis … it was too urgent, we just had to do things. People use the phrase "balls to the wall," and it was very telling because there was this focus on like we have to do everything, we have to save everyone and we don't have time to figure out if people are getting paid or if their health insurance is kicking in for their jobs or these petty bourgeois concerns, we have to … be organizing and responding to all these crises. And the fact that we were building an organization that was completely unsustainable … raising those concerns was put down in a very gendered way.… It's like whatever landed on fire in our laps that's what we responded to.

Deciding what counts as fire when it lands in your lap is a gendered process. There are two dimensions here. First, urgency is socially constructed, a designation based on priorities and meanings which are always already gendered. Rosana observed that

> there was also this very macho idea of organizing where it was about results.… you have to have certain results and it has to be sexy to funders and it has to have certain media attention. And so ultimately to me it was gendered.

Experiencing wage theft is urgent, experiencing sexual assault is not; being unable to return home because of flooded housing is urgent, being unable to return home because of caregiving responsibilities is not; having scheduling flexibility and being able to work all night is a priority in an urgent situation, while knowing the city, speaking Spanish and having feminist credentials are not.

Second, urgency encourages certain kinds of practices that are justified by expediency. Several organizers described modalities of what they called patriarchal, authoritarian organizing that were rationalized by crisis and scarcity. These modalities do not have time for feminist principles or concerns, as Rosana noted above, such as the inclusion of gender as politically important, attentiveness to default conduits of power and leadership, sustainable models of balanced or holistic organizing, or an ethic of care and justice for organizers.

Kai deepens this analysis by explaining that authoritarian modalities are resistant to "facilitative organizing," which she describes as less patriarchal and more deliberative and reflective. Facilitative organizing, she explains, means organizers

> will lay groundwork, that we will engage people in a range of different activities to stimulate critical thinking, and that we will provide resources so that folks can engage their own critical production.… But does that work in emergencies, does

that work when there is crisis at work, crisis upon crisis upon crisis that we're constantly battling?

Khalil similarly distinguished organizing modalities:

> you see the sort of gender dynamics play out in the way that ... it's always about this sort of growth of power in opposition to an oppressor as opposed to – using gendered terms – nurturing and developing an alternative livelihood, an alternative society....What I was really hoping for, like out of the neighborhood planning process...that there would be this sort of process of actual organizing of communities, like permanently organizing communities into a local organized institutional basis.

Engendering is always intersectional; gendered modalities were produced and consumed in racialized ways. Bay Love, a young White man who had just graduated from a liberal arts college, framed it this way:

> [It] just felt like the gender dynamics—but really like the patriarchy that was present there—was just really really palpable because it was like, post-Katrina, we got to carry guns, we got to get water, we got to bust through the piles of cars and the police oppression to get there. It just felt like it was really important to be big and strong and manly, and [an older Black male leader] is a figurehead, in my mind, so sort of masculine, like really big and has a really deep voice and [a young White man who was later exposed as an FBI informant] was a lot of that too you know. A lot of the sort of militancy that was around ... at that time was all very sort of gendered and male-dominated.... I felt like I was sort of proud ... to be able to say we were part of an organization where people would carry guns if they needed to. [Pause] Even though I never really wanted to myself.

Beyond gendered bodies

A final way in which the valorization and centralization of masculinity functioned is expressed in the advancement of certain women leaders in the movement over others. Several prominent, middle-aged, local Black women were widely recognized as movement leaders. Some had professional credentials, such as law degrees, where most of the senior male leadership were authorized instead by political experience. In each case, the senior woman led with an explicit race and class framework, rarely mentioning gender. It seemed that women who rose more easily to prominence in the movement were those who promoted gender-neutral politics; their political priorities and commitments more closely resembled the men's than they did those of feminist women in the movement who were seen as having "special" interests. When charges of sexism arose, people were quick to point to the women's leadership as counterevidence, focusing on the gendered body count and not on gendered practices or feminist principles of social justice organizing.

Challenges to men and masculinity: consciousness-raising opportunities

While most of the gender and disaster literature indicates that gender is more likely to be aggressively reproduced in times of crisis, there is evidence that the radical if temporary alteration of daily life provides opportunity for gender power subversion. In the context of the Movement for a Just Reconstruction after Katrina, gendered power dynamics were disrupted in two primary ways. The first was the transformation of some women into organizers. The second was feminist consciousness-raising. I focus here on the latter.

Endesha Juakali is a middle-aged Black man who grew up in New Orleans public housing and spent much of his life working with and for public housing residents. Before Katrina he did not read gender as central to "day to day struggles," instead focusing on race and class. This changed after the disaster:

> Everything I've began [sic] to challenge on this issue has been post-Katrina.... if you would have talked to me prior to Katrina you probably wouldn't still be sitting here because I was strictly nationalist, macho, kind of, "we don't have time to worry about that, that's personal, we need to stick with fighting the ... revolutions around race and class.... that's an issue that just diverts us from the real issues." So I mean yeah, most of my growth on this area has come since Katrina.

He was in a process that was driven by explicit conversations that happened organically in the midst of post-Katrina organizing, noting, "I've been called out quite a few times." He describes one particular incident:

> We were at a ... housing action group meeting.... And this brother did this rap thing and ... he called Bush a faggot and I mean you had the people who was really upset about him using the word and I was like well, we need to stick to the workshop. That's going to divert us.... So then they had to stop the whole session to talk about why I didn't think it was important.... [T]hey had quite a few lesbian and gays and different people who was like, that's important to us kind of thing, it's a part of the work. I mean we can't work with you if you don't respect.

These exchanges influenced Endesha to incorporate gender into his political framework:

> We're all growing and the more I worked with people who bring forth the issue—if there's a contradiction between racism and classism and getting rid of class would not resolve the race issue—well I'm feeling the same thing now [about gender]. Getting rid of race and class will not resolve the gender issue, that's an issue that's going to have to be dealt with just pretty much like racism.

Endesha believed that the Movement for a Just Reconstruction had exposed him to feminist concerns in a new way: "[Normally] I just don't think about those things.

Well, let me say this, this gender issue was – I probably never mentioned gender prior to 2005." The political landscape of New Orleans before Katrina, as with most US cities, was composed of relatively siloed political communities. Katrina loosed the boundaries in fruitful ways creating a new context for organizers kept apart by political differences and the demands of daily life. For a short time, Katrina was a new meta-frame that sought to link issues and people generally understood to be incompatible. While it did not often prevail, there was at times a sense among organizers that the struggle for justice after the devastation required new levels of solidarity.

During this window, there were opportunities for what Kai described as a political stance of "opening up." Endesha observed:

> Yeah, now I do absolutely [consider gender] because it affects the struggle. I mean not only in regards to a male and female but the whole lesbian, bisexual, transgender, all of that. I've experienced very good, wonderful comrades in these areas who think these areas are very important to them and the same way that I would expect they would understand how racism affects me I should work to understand how it affects those comrades.

After the peak period of the movement and many feminist attempts to make it more feminist, several male organizers expressed remorse that they had not supported this effort. Chris Crass, a White antiracist, feminist, nonlocal organizer who mentored young White activists reflected on his group's hesitation to be explicitly feminist:

> we have an understanding organizationally around gender and around patriarchy and around feminism, but I think we've struggled around how to articulate that.… I think the ways that we've … talked about things had really led to people deemphasizing gender, deemphasizing the negative impacts of patriarchy on female-socialized people, the movement, on gender dynamics within organizations.

Jordan had a similar regret. He recalled the period in 2006 when women activist-volunteers were sexually assaulted by male activist-volunteers in Common Ground, one of the movement groups:

> I wish I had been thinking more about how gender stuff was playing out. For example, within Common Ground, the stuff with sexual assault.… I did not see that stuff coming. I think that more of my analysis was that White males … were coming to New Orleans with this racist, colonial, imperial attitude. So … I felt like White males were a point to reach out to but I felt like that they needed to be especially reached out to on a race-level, not on a gender-level.

Despite the fact that feminist principles never became central to the movement, a fact that likely undermined its efficacy and its accomplishment of racial and economic justice (Luft, forthcoming), there are signs that they had an impact. As a result of the determination of feminist activists, some organizers began to develop a gender analysis while others recommitted to bringing it more centrally into their work.

Conclusion

Major disaster burns, disintegrates or, in the case of Hurricane Katrina, washes away material and social structures of daily life. Despite the power of these destructive forces, however, gender usually remains. Normative structural and cultural patterns that promote men and masculinity interact with the exceptional conditions of disaster to produce common gendered outcomes in abnormal contexts. Less often, but important, are the ways in which disaster catalyzes new possibilities. In the case of the social movement after Katrina, activists and organizers from different political orientations came together to fight for a just reconstruction, and in the process some experienced feminist consciousness-raising.

The point of this chapter is not to condemn men for being movement leaders in a time of crisis. Indeed, most of those in the Movement for a Just Reconstruction were visionary, self-sacrificing and deeply committed to the struggle for justice in what were traumatizing, disorienting and challenging conditions. Instead, my aim has been to shed light on the ways in which gender was part of the disastrous landscape. Normative masculinity was often the default or ruling relation in the movement. This social fact had consequences not only for women and for feminist principles, but also for ostensibly gender-neutral practices, such as organizing strategy and tactics, and therefore for outcomes. While the coincidence of men and masculinity means that promoting differently gendered people into positions of movement leadership is likely to make a difference, the gender of bodies alone is not sufficient. Instead, identifying and interrupting masculinity in values and practices coded gender neutral is necessary to decenter masculinity and promote feminist, intersectional outcomes.

Note

1   With gratitude to the organizers of the post-Katrina Movement for a Justice Reconstruction in New Orleans. I am particularly indebted to Shana griffin, who gets credit for the best of what is in this chapter.

References

Connell, R.W. and Messerschmidt, J. (2005) Hegemonic masculinity: Rethinking the concept. *Gender & Society*, 19(6), pp. 829–859.

Enarson, E. (2012) *Women confronting natural disaster: From vulnerability to resilience.* Boulder, CO: Lynne Rienner Publishers.

Luft, R. (2008) Looking for common ground: Relief work in post-Katrina New Orleans as an American parable of race and gender violence. *National Women's Studies Association Journal*, 20(3), pp. 5–31.

Luft, R. (forthcoming) Racialized disaster patriarchy: An intersectional model for understanding disaster ten years after Hurricane Katrina. *Feminist Formations*.

Pacholok, S. (2013) *Into the fire: Disaster and the remaking of gender.* Toronto: University of Toronto Press.

Williams, C.L. (2013) The glass escalator, revisited: Gender inequality in neoliberal times. *Gender & Society*, 27(5), pp. 609–629.

4 Hyper-masculinity and disaster

The reconstruction of hegemonic masculinity in the wake of calamity

Duke W. Austin

Although the data are limited, numerous qualitative and mixed-method studies indicate that gendered sexual and domestic violence often increase in the aftermath of disasters (Fothergill, 1996; Enarson and Morrow, 1998; Enarson *et al.*, 2007). Reports show that violence against women increased following the Loma Prieta earthquake, Hurricane Andrew and the Grand Forks flood, all in the United States (Laudisio, 1993; and see Enarson, 2012 for a review). In addition, research indicates that gendered violence increased following Hurricane Mitch in Nicaragua (Cupples, 2007), the Indian Ocean tsunami in Sri Lanka (Fisher, 2010), the Whakatan flood and the Canterbury snowstorm in New Zealand (Houghton, 2009; Houghton *et al.*, 2010), the Sichuan earthquake in China (Chan and Zhang, 2011) and the 2010 Haiti earthquake (Bell, 2010), among others.

A common misconception is that heightened levels of stress following a disaster lead to increases in gendered violence (Enarson, 2012). However, research has been unable to establish a direct positive relationship between disaster-related stress and domestic violence (Clemens *et al.*, 1999). In contrast to individual-level analyses, I focus on structural causes of post-disaster gendered violence. While individual-level analysis has an important role to play in uncovering the causes of gendered violence following a disaster, focusing only on the individual fails to account for structural changes that can lead to increased levels of violence.

In this chapter, I theorize that the increase in men's violence against women is the result of a breakdown and reconstruction of masculinity following the impairment of the institutional structures that support hegemonic masculinity. During routine, non-disaster times, hegemonic masculinity maintains dominance less through actual violence and more through the threat of violence. This less violent form of dominance depends, however, on the orderly maintenance of institutional structures, and natural disasters often disrupt those structures. As a result, disaster masculinity, a form of hyper-masculinity, emerges post-disaster. Disaster masculinity often leads to increased levels of gendered violence.

The gendered violence that takes place following disasters must be understood in the context of gendered violence that occurs during routine, non-disaster times. Violence against women during these routine times occurs as a result of the subordinate status of women relative to men. Men's violence serves to express and reinforce the gendered hierarchy while controlling and maintaining men's power, privilege and dominance (Bunch, 1990; Carillo, 1993; Kaufman, 1999; Fisher, 2010). As such, I begin with a discussion of how masculinity is constructed in non-disaster patriarchal society. Then, I examine the ways in which a particular disaster, Hurricane Katrina, degraded the institutions that support the constructions of masculinity. Finally, I theorize how disaster masculinity, a form of hypermasculinity, emerges from the structural loss created by a disaster, which often leads to increased levels of gendered violence.

Routine construction of masculinity

Gender is socially constructed, but also a matter of choice, which requires a pre-choice body. While strict social constructionists disagree with the notion of a preexisting body, choice and the social construction of gender can be compatible. Distant subjects do not choose gender at a particular moment. Instead, the individual creates and reifies it in interaction with the larger social context. That is not to say, however, that genders are sets of alternative lifestyles that an individual browses like consumers in a market. Instead, there exist hard compulsions under which gender configurations are formed (Connell, 1995; Salih and Butler, 2004). Individuals constantly renegotiate these gendered identities in their social lives (Jackson, 1991; Hague, 1997).

Multiple masculinities

Gender does not operate in society apart from other structures in society. Instead, gender interacts with race, class, age, sexuality and a host of other categories to create multiple femininities and masculinities (Hill-Collins, 2005). In this chapter, I risk oversimplification by only discussing two masculinities that operate within the milieu of social structures—namely hegemonic and marginalized masculinities. Hegemony, originally used in discourses of class dynamics, refers to the cultural dynamics by which a group claims and sustains a leading position in society. As Connell writes,

> Hegemonic masculinity can be defined as the configuration of gender practice which embodies the currently accepted answer to the problem of the legitimacy of patriarchy, which guarantees (or is taken to guarantee) the dominant position of men and the subordinate position of women.
>
> (1995, p. 77)

Hegemonic masculinity includes the ability to control and make decisions, to act rationally, to control emotions and to exhibit physical courage and muscular prowess (Jackson, 1991). Hegemonic masculinity is not, however, immutable—always and everywhere the same. Instead, it is whichever masculinity occupies the hegemonic position within a specific social context. In other words, hegemonic masculinity embodies a currently accepted strategy. When the social context changes, the bases that authorize the hegemony of masculinity erode, and new groups may challenge old authority and construct new hegemony. Masculinity, nonetheless, only becomes hegemonic when the cultural ideal and institutional power correspond on the collective level and grant masculinity the authority to dominate (Connell, 1995).

In addition to interacting between femininities and masculinities, social categories such as race, class, sexuality, etc. also form integral parts of the dynamic between masculinities themselves. For example, masculinities can be marginalized along lines of class and race while dominating in the arena of gender. Working-class men hold marginalized positions in comparison with middle-class men. Similarly, men of color in the United States hold inferior societal positions in relation to white men (Staples, 2004; Hill-Collins, 2005).

Still, marginalized masculinities play symbolic roles for the construction of hegemonic masculinity. For example, working-class masculinities are exemplars of masculine toughness, and black sports stars are exemplars of the physical embodiment of masculinity. As such, marginalized masculinity and hegemonic masculinity are interrelated. The first is always relative to the authorization of the second (Connell, 1995; Hill-Collins, 2005).

Hegemonic masculinity among white men sustains the institutional oppression that has framed the construction of masculinities in communities of color. For example, Robert Staples (2004) argues that Black men are in conflict with hegemonic definitions of masculinity in US culture, definitions that imply autonomy and mastery of one's environment and the accumulation of wealth. Working-class men and men of color face similar alienation from hegemonic masculinity. As such, working-class men of color have to confront contradictions between the hegemonic expectations attached to being male and the proscriptions on behavior and achievement of goals. Such men are thereby stigmatized for failing to live up to the standards of masculinity on the one hand and for being too "tough" or too "physical" on the other (Staples, 2004).

Marginalized masculinities, including the forms of masculinity available to working-class men and men of color, create a feeling of emasculation and powerlessness in the arenas of class and race, even though dominance is maintained in the arena of gender. For example, bell hooks (1981) recognizes the way black male political leaders of the civil rights movement, while fighting racial oppression, expected Black women to assume a subservient role. As hooks asserts,

> [They] made no effort to emphasize that patriarchal power, the power men use to dominate women, is not just the privilege of upper- and middle-class white men, but the privilege of all men in our society regardless of their race or class.
>
> (p. 87)

Marginalized men want access to the power of hegemonic masculinity but have been unable to obtain it. These men often attempt to display their masculinity in other ways, specifically by forcing marginalized women into subservient roles (hooks, 1981).

The relationship between hegemonic and marginalized masculinities is not a fixed formation but rather a configuration of practice that is generated in a specific social context. The relationships between the two masculinities and between masculinities and femininities change as the social context changes, allowing us to analyze specific masculinities (Connell, 1995).

Hyper-masculinity exaggerates the toughness, violence and terror of hegemonic and marginalized masculinities. As men experience a sense of loss of dominance in social institutions such as the state, the economy and the family, they may resort to hyper-masculinity in an effort to restore their hegemonic influence. As I demonstrate below, both hegemonic and marginalized masculinities play a role in the construction of hyper-masculinity post-disaster.

Hegemonic masculinity and violence against women

Interpersonal, gendered violence is often used to create authority and often underpins and supports authority. Men with marginalized masculinities may resort to violence to establish dominance in their relationships and to mimic the power in hegemonic masculinity. Men with hegemonic masculinities may resort to gendered violence when they feel their hegemony fading. However, hegemonic masculinity is defined more by its successful claim to authority than its use of direct violence (Connell, 1995).

Levels of men's violence against women are alarming during non-disaster, "normal" times. Violence against women is one of the most pervasive forms of human rights abuse worldwide, and the perpetrators of violence against women are most often their intimate male partners (Center for Social Development, 2010). Gendered violence denies women their human rights, prevents them from full participation in society and keeps them from achieving their full potential (Bunch, 1990; Carillo, 1993; Heise *et al.*, 1999).

Social structures such as the family, the workplace, religion and the state support hegemonic masculinity. As such, hegemonic masculinity relies less on violence and more on the status quo to maintain the gendered hierarchy during routine, non-disaster times. In the following section, I provide an example of how the degradation of social structures might lead to an expression of hyper-masculinity following a large disaster.

Hurricane Katrina, the structural supports of masculinity, and men's violence against women

In the US, one of the better-known examples of increased violence against women happened following Hurricane Katrina. According to a study by the National Sexual Violence Resource Center, 47 cases of sexual assault were reported in the immediate aftermath of Hurricane Katrina, and reported cases of sexual assault in New Orleans increased 45% in the six months following the storm. Results show that 93% of the perpetrators were male, with the remainder unspecified, and 93% of the victims were female (National Sexual Violence Resource Center, 2006;

Thornton and Voigt, 2007). Existing survivors of domestic violence often experienced heightened levels of gendered violence during and after the storm (Jenkins and Phillips, 2008), and the city of New Orleans saw a 68% spike in the reports of rape (McCarthy and Philbin, 2007, cited in Jenkins and Phillips, 2008).

The aftermath of Hurricane Katrina saw the degradation of three social structures that support hegemonic masculinity—the state, the economy and the institution of family.

Hurricane Katrina and the state

National polls reflect widespread criticism of the government's response to Hurricane Katrina, including the local, state and federal government (CNN, 2005; Langer, 2005). Following the storm, many public services were disrupted or altogether unavailable in the Gulf Coast. Storm victims lacked access to schools, childcare, hospitals, courts, police, transportation, transition centers, counselors, crisis lines and even potable water (Laska and Morrow, 2006; Jenkins and Phillips, 2008).

While this lack of public services decreased the ability for survivors of gendered violence to seek protection from their abusers (Enarson, 2012), it may have simultaneously given men the sense that they no longer had control or dominance in their lives. In the absence of police service, some men chose violence and threats of violence to reassert or achieve a feeling of control. Indeed, many hurricane survivors, a majority of whom were men, felt a need to arm themselves during the disruption of public services in order to protect themselves, their families and their property (Miller and Rivera, 2007).

At the same time that public services were curtailed or unavailable, the government imposed an authoritarian state with strict curfews and National Guard soldiers. Authorities harassed residents and denied services to people without social security numbers (Ross, 2012). Soldiers patrolled the streets, the emergency shelters and even their sleeping spaces (Austin and Miles, 2006). In some cases, the state authorities were guilty of perpetrating violence themselves (INCITE! 2012; Greene, 2015). Undoubtedly, these state-sponsored institutions held immense control over the lives of hurricane survivors.

On one hand, the disruption of public services contributed to social disorganization and a loss of order in the lives of individuals. On the other hand, state institutions such as the National Guard diminished individual control. With the weakening of social order and stability in the lives of everyday people and with the decline of individual control, it is possible that some men chose hyper-masculine violence in order to regain or achieve a sense of masculine control, order and dominance.

Hurricane Katrina and the economy

Estimates of property damage during Hurricane Katrina topped $125 billion (Associated Press, 2005; Knabb *et al.*, 2006), making it the most costly disaster in US history. Local businesses also faced unprecedented difficulties in resuming operations. Businesses not only suffered from the physical damage to their structures, but also from the loss of operations while public infrastructure was inaccessible, the dispersion of employees and the dislocation of the customer base (Boin and McConnell, 2007; Corey and Deitch, 2011). All told, approximately 180,000 people lost their jobs following Hurricane Katrina (Griffin and Woods, 2009).

When the economy suffers and people lose their jobs, they lose their ability to provide for themselves and their loved ones. They can't pay rent or make mortgage payments, and they can't buy food and other necessities. Without employment, people lose the power to control their own lives. While both women and men experience this loss of power and control, men are especially unable to fulfill the expectations of their socially constructed masculine gender identities. Following Hurricane Katrina, some men used hyper-masculine violence, and it is possible they did so to reassert the sense of dominance that had been previously supported by the economic structure.

Hurricane Katrina and the institution of family

Hurricane Katrina disrupted the institution of family as well. Sadly, at least 1,833 people lost their lives because of Katrina (Associated Press, 2005; Knabb *et al.*, 2006), and an estimated 300,000 homes were destroyed (Department of Homeland Security, 2006). In addition to the turmoil that the hurricane brought to families and friends who lost loved ones, it severely disrupted social ties among the survivors. The disaster separated immediate family members and prevented survivors from contacting extended family and friends (Weems *et al.*, 2007).

While many close and intimate relationships provided positive emotional support in the aftermath of the storm, many others contributed to strain, stress, and discord (Reid and Reczek, 2011). As many as one-third of married adults in the affected area separated in the year following Katrina, a number that is more than twice the national rate (Rendall, 2011). Given the housing shortage following the disaster, many families had to live with extended family members, double up with other families, or live in cramped FEMA trailers.

These changes to the family structure lead to discord and disorganization in one's life. Marginalized and privileged men who are accustomed to the greater levels of power and control in their families may choose hyper-masculine violence as a means to reassert a sense of hegemony during periods of social disorganization such as the Hurricane Katrina disaster.

Large disasters such as Hurricane Katrina can lead to the perceived loss of masculine hegemony in the state, in the economy, and in the institution of family. In the following section, I theorize how this loss of dominance can lead to the emergence of disaster masculinity and increased levels of gendered violence.

Disaster masculinity: structural disorganization and the increase in hyper-masculinity

Social disorganization theory derives from the conflict perspective of violence, and highlights the "inability of a community structure to realize the common values of its residents and maintain effective social controls" (Shaw and McKay, 1942; Sampson and Groves, 1989; Sampson and Wilson, 2005, p. 182). The theory states that geographic zones can experience increases in nongendered violence when those zones experience rapid transition and physical deterioration. Disaster researchers have applied the theory to the study of sexual violence among internally displaced women following Hurricane Katrina (Fagen *et al.*, 2011). Following them, I theorize that social disorganization can lead to reconstructions of masculinity, which often results in the emergence of a violent form of hyper-masculinity.

Masculinities are not intrinsically powerful (Hague, 1997). Instead, hegemony depends on contextual and circumstantial factors such as the structure of the state, the economy, and the family. When something like a natural disaster degrades those structures, and men feel their hegemony is in crisis, they respond differently. Arguably, some men feel inadequate to fulfill the expectations of their gender, and they make the choice to act in violent ways.

The feeling of inadequacy transcends hegemonic and marginalized distinctions, though is likely to be especially strong for marginalized men who are already challenged in a society structured by class, racial, and other inequalities. Feelings of inadequacy build in men, increasing the numbers of men who to choose to act in violent, abusive ways derived from a sense of hyper-masculine masculinity.

True, many social structures provide men *more* power and control than women in the response and recovery to a disaster. For example, girls are more likely than boys to be removed from school post-disaster, many economic relief programs mainly benefit men, and men often experience increased power in family decisions such as evacuating and relocating. As such, women tend to lose more power and control than men following a large disaster (Enarson, 2012). Even so, men are likely to perceive a loss of dominance relative to their own position prior to the disaster, not in relation to the greater loss that women suffer following a disaster. In addition, men are more likely than women to feel a sense of inadequacy given the dominant expectations of men's socially constructed gender identity. While not all men chose hyper-masculine violence as a means to reassert hegemony during social disorganization, far too many men do. The fact that more men do not engage in gendered violence following a disaster may speak to the enduring resilience of masculine hegemony.

Disaster masculinity

Rachel Luft uses the term "disaster masculinity" to describe a form of hyper-masculinity that arose in a New Orleans volunteer community following Hurricane Katrina. For Luft, disaster masculinity enabled the gendered violence that took place within the volunteer community and is marked by white colonial and neo-liberal features, including "intimacy, racialized masculinity, authoritarianism, adventure, and paternalism" (2008, p. 22). Here, I expand her description of disaster masculinity to include the form of hyper-masculinity that emerges post-disaster among men with both hegemonic and marginalized masculinities. Some men may choose to engage in gendered violence following a disaster in order to reassert a perceived loss of power and control as the social structures that support masculinity become disorganized.

The structural breakdown and disorganization following a disaster can exacerbate the excuses and justifications that some men give for gendered violence during non-disaster, "normal" times. Tim Beneke interviewed convicted rapists about the motivation for their crimes. The following quote, borrowed from his article "Men on Rape" (2013), demonstrates how some men justify the use gendered violence to regain a sense of masculine dominance:

> As a man, you're taught that men are more powerful than women, and that men always have the upper hand, and that it's a man's society.
>
> In this society, if you ever sit down and realize how manipulated you really are it makes you pissed off—it makes you want to take control... [Women] are a very easy target because they're out walking along the streets, so you can just grab one and say, 'Listen, you are going to do what I want you to do,' and it's an act of revenge against the way you've been manipulated.
>
> (Beneke, 2013, p. 568)

Following a disaster, hyper-masculinity emerges as a choice. Even more than during non-disaster times, men are likely to feel a loss of power or control because they are unable to live up to the expectations of their socially constructed gender identities. The crisis situation provokes attempts to restore hegemonic masculinity. As a result, some men resort to hyper-masculinity and gendered violence when they fear a loss of dominance in social structures such as the state, the economy, or the institution of family.

Conclusion

During routine, non-disaster times, masculinities are constructed to have dominance over femininity. The social institutions of the state, the economy, and the institution of family support that dominance. Inter-personal, gendered violence is often used as a tool of masculinities, both to construct hegemony and to defend it. However, disasters often cause a breakdown of the social institutions that support masculine dominance. The structural supports discussed here—the state, the economy, and the institution of family—are weakened, and the hegemony of masculinity is threatened. Disaster masculinity emerges, and some men resort to gendered violence in an effort to achieve or restore the dominance of hegemonic masculinity.

References

Associated Press (2005) Katrina damage estimate hits $125B. *The Associated Press*, September. Available from http://www.usatoday.com/money/economy/2005-09-09-katrina-damage_x.htm (accessed May 3, 2007).

Austin, D.W. and Miles, M. (2006) Crisis in black and white: Katrina, Rita, and the construction of reality. In: National Hazards Centre (eds.) *Learning from catastrophe:*

Quick response research in the wake of Hurricane Katrina. Boulder, CO: University of Colorado, pp. 151–173.

Bell, B. (2010) Haiti's women: "our bodies are shaking now." *Common Dreams*. Available from http://www.commondreams.org/views/2010/03/25/haitis-women-our-bodies-are-shaking-now (accessed July 18, 2015).

Beneke, T. (2013) Men on rape. In: Kimmel, M.S. and Messner, M.A. (eds.) *Men's Lives*, 9th edn. Upper Saddle River, NJ: Pearson Education, pp. 563–568.

Boin, A., and McConnell, A. (2007) Preparing for critical breakdowns: The limits of crisis management and the need for resilience. *Journal of Contingencies and Crisis Management*, 15(1), pp. 50–59.

Bunch, C. (1990) Women's rights as human rights: Toward a re-vision of human rights. *Human Rights Quarterly*, 12(4), pp. 486–498.

Carillo, R. (1993) Violence against women: An obstacle to development. In: Turshen, M. and Holcomb, B. (eds.) *Women's lives and public policy: The international experience*. Portsmouth: Greenwood Publishing Group, pp. 99–113.

Center for Social Development (2010) *The world's women: Trends and statistics*. New York: United Nations.

Chan, K.L. and Zhang, Y. (2011) Female victimization and intimate partner violence after the May 12, 2008, Sichuan Earthquake. *Violence and Victims*, 26(3), pp. 364–376.

Clemens, P., Hietala, J., Rytter, M., Schmidt, R. and Reese, D. (1999) Risk of domestic violence after flood impact: Effects of social support, age, and history of domestic violence. *Applied Behavioral Science Review*, 7(2), pp. 199–206.

CNN (2005) Poll: Most Americans believe New Orleans will never recover: Still, 63 percent of respondents say city should rebuild. *CNN.com*, September 8. Available from http://www.cnn.com/2005/US/09/07/katrina.poll/index.html (accessed May 3, 2007).

Connell, R.W. (1995) *Masculinities*. Berkeley, CA: University of California Press.

Corey, C.M. and Deitch, E.A. (2011) Factors affecting business recovery immediately after Hurricane Katrina. *Journal of Contingencies and Crisis Management*, 19(3), pp. 169–181.

Cupples, J. (2007), Gender and Hurricane Mitch: Reconstructing subjectivities after disaster. *Disasters*, 31(2), pp. 155–175.

Department of Homeland Security (2006) Statement for Secretary Michael Chertoff, U.S. Department of Homeland Security, before the Senate Committee on Homeland Security and Governmental Affairs. Presentation given to the U.S. Senate on February 14, 2006. Available from http://72.14.253.104/search?q=cache:2XHIasDWu6cJ:hsgac.senate.gov/_files/021406Chertoff.pdf+300,000+homes+destroyed&hl=en&ct=clnk&cd=4&gl=us&client=safari (accessed May 3, 2007).

Enarson, E. (1997) Responding to domestic violence and disaster: Guidelines for women's services and disaster practitioners. *BC Institute Against Family Violence*. Available from http://www.gdnonline.org/resources/dv-and-disaster.doc (accessed September 23, 2015).

Enarson, E. (1998) Through women's eyes: A gendered research agenda for disaster social science. *Disasters*, 22(2), pp. 157–173.

Enarson, E. (2012) *Women confronting natural disaster: From vulnerability to resilience*. Boulder, CO: Lynne Rienner Publishers.

Enarson, E., Fothergill, A. and Peek, L. (2007) Gender and disaster: Foundations and directions. In: Rodriguez, H., Quarantelli, E.L. and Dynes, R.R. (eds.) *Handbook of disaster research*. New York: Springer, pp. 130–146.

Enarson, E. and Morrow, B.H. (1998) *The gendered terrain of disaster*. New York: Praeger.

Fagen, J.L., Sorensen, W. and Anderson, P.B. (2011) Why not the University of New Orleans? Social disorganization and sexual violence among internally displaced women of Hurricane Katrina. *Journal of Community Health*, 36(5), pp. 721–727.

Fisher, S. (2010) Violence against women and natural disasters: Findings from post-tsunami Sri Lanka. *Violence against Women*, 16(8), pp. 902–918.

Fothergill, A. (1996) Gender, risk, and disaster. *International Journal of Mass Emergencies and Disasters*, 14(1), pp. 33–56.

Greene, R. (2015) *Shots on the bridge: Police violence and cover-up in the wake of Katrina*. Boston, MA: Beacon Press.

Griffin, S. and Woods, C. (2009) The politics of reproductive violence. *American Quarterly*, 61(3), pp. 583–591.

Hague, E. (1997) Rape, power and masculinity: The construction of gender and national identities in the war in Bosnia-Herzegovina. In: Lentin, R. (ed.) *Gender and catastrophe*. New York: Zed Books, pp. 50–63.

Heise, L., Ellsberg, M. and Gottemoeller, M. (1999) Ending violence against women. *Population Reports, Series L: Issues in World Health*, 11, pp. 1–43.

Hill-Collins, P. (2005) *Black sexual politics*. New York: Routledge.

hooks, b. (1981) *Ain't I a woman?* Boston: South End Press.

Houghton, R. (2009) "Everything became a struggle, absolute struggle": Post-flood increases in domestic violence in New Zealand. In: Enarson, E. and Chakrabarti, P.G.D. (eds.) *Women, gender and disaster: Global issues and initiatives*. New Delhi: Sage, pp. 99–112.

Houghton, R., Wilson, T., Smith, W. and Johnson, D. (2010) "If there was a dire emergency, we never would have been able to get in there": Domestic violence reporting and disasters. *International Journal of Mass Emergencies and Disasters*, 28(2), pp. 270–293.

INCITE! Women of Color Against Violence (2012) INCITE! Statement on Katrina. In: David, E. and Enarson, E. (eds.) *The women of Katrina: How gender, race, and class matter in an American disaster*. Nashville, TN: Vanderbilt University Press, pp. 3–6.

Jackson, P. (1991) The cultural politics of masculinity: Towards a social geography. *Transactions of the Institute of British Geographers*, 16(2), pp. 199–213.

Jenkins, P. and Phillips, B. (2008) Battered women, catastrophe, and the context of safety after Hurricane Katrina. *NWSA Journal*, 20(3), pp. 49–68.

Kaufman, M. (1999) The seven p's of men's violence. Available from: http://www.michaelkaufman.com/1999/the-7-ps-of-mens-violence/ (accessed December 12, 2015).

Knabb, R.D., Rhome, J.R. and Brown, D.P. (2006) Tropical cyclone report, Hurricane Katrina, 23–30 August 2005. Miami, FL: National Hurricane Center, National Oceanic & Atmospheric Administration (NOAA), U.S. Department of Commerce.

Langer, G. (2005) Poll: Bush not taking brunt of Katrina criticism: Hurricane preparedness is faulted; fewer blame Bush for problems. *ABC News*, September 12. Available from http://abcnews.go.com/US/HurricaneKatrina/story?id=1094262 (accessed September 23, 2015).

Laska, S. and Morrow, B.H. (2006) Social vulnerabilities and Hurricane Katrina: An unnatural disaster in New Orleans. *Marine Technology Society Journal*, 40(4), pp. 16–26.

Laudisio, G. (1993) Disaster aftermath: Redefining response—Hurricane Andrew's impact on I&R. *Alliance of Information and Referral Systems*, 15, pp. 13–32.

Luft, R.E. (2008) Looking for common ground: Relief work in post-Katrina New Orleans as an American parable of race and gender violence. *NWSA Journal*, 20(3), pp. 5–31.

McCarthy, B. and Philbin, W. (2007) City's violent crime up in third quarter. *The Times Picayune*, Metro section B: 1 and 2.

Miller, S.M. and Rivera, J.D. (2007) Landscapes of disaster and place orientation in the aftermath of Hurricane Katrina. In: Brunsmaa, D.L., Overfelt, D. and Picou, J.S. (eds.) *The sociology of Katrina: Perspectives on a modern catastrophe*. Lanham, MD: Rowman & Littlefield.

National Sexual Violence Resource Center (2006) *Hurricanes Katrina/Rita and sexual violence: Report on database of sexual violence prevalence and incidence related to Hurricanes Katrina and Rita*. Available from http://www.nsvrc.org/publications/nsvrc-publications/hurricanes-katrinarita-and-sexual-violence-report-database-sexual-vi (accessed on July 18, 2015).

Reid, M. and Reczek, C. (2011) Stress and support in family relationships after Hurricane Katrina. *Journal of Family Issues*, 32(10), pp. 1397–1418.

Rendall, M.S. (2011) Breakup of New Orleans households after Hurricane Katrina. *Journal of Marriage and Family*, 73(3), pp. 654–668.

Ross, L.J. (2012) A feminist perspective on Katrina. In: David, E., and Enarson, E. (eds.) *The women of Katrina: How gender, race, and class matter in an American disaster*. Nashville, TN: Vanderbilt University Press, pp. 15–23.

Salih, S. and Butler, J. (2004) *The Judith Butler reader*. Malden: Blackwell.

Sampson, R.J. and Groves, W.B. (1989) Community structure and crime: Testing social-disorganization theory. *American Journal of Sociology*, 94(4), pp. 774–802.

Sampson, R.J. and Wilson, W.J. (2005) Toward a theory of race, crime, and urban inequality. In: Gabbidon, S., and Greene, H.T (eds.) *Race, crime, and justice*. New York and London: Routledge, pp. 37–54.

Shaw, C.R. and McKay, H.D. (1942) *Juvenile delinquency and urban areas*. Chicago, IL: Routledge.

Staples, R. (2004) Stereotypes of black male sexuality: The facts behind the myths. In: Kimmel, M. and Messner, M. (eds.) *Men's Lives*, 6th edn. Boston, MA: Pearson Education, Inc., pp. 375–380.

Thornton, W.E. and Voigt, L. (2007) Disaster rape: Vulnerability of women to sexual assaults during Hurricane Katrina. *Journal of Public Management and Social Policy*, 13(2), pp. 23–49.

Weems, C.F., Watts, S.E, Marsee, M.A., Taylor, L.K., Costa, N.M, Cannon, M.F, Carrion, V.G. and Pina, A.A. (2007) The psychosocial impact of Hurricane Katrina: Contextual differences in psychological symptoms, social support, and discrimination. *Behaviour Research and Therapy*, 45(10), pp. 2295–2306.

5 Rereading gender and patriarchy through a "lens of masculinity"

The "known" story and new narratives from post-Mitch Nicaragua

Sarah Bradshaw

In 1998, Hurricane Mitch brought strong winds and severe flooding and in just five days an entire year's rainfall was recorded in southwestern Honduras and the Pacific coastlines of Nicaragua and El Salvador.[1] It is estimated that the hurricane had a direct impact on one in every ten people across the region, and caused damage totaling 4 billion dollars to the productive sector. Mitch was the fourth most intense hurricane in the Atlantic in the twentieth century, and also the first in the region to be explored as a gendered event. The external understanding of the gendered experience of Hurricane Mitch in Nicaragua was constructed by a small number of researchers, including myself (see Delaney and Shrader, 2000; CIETinternational, 1999, 2001; Bradshaw, 2001a; Cupples, 2007). As one of the first gendered readings of such an event it was important that women's voices were privileged; and while these studies discussed both women and men, it is women who are put at the center of the analysis and men explored in relation to them.

A popular thesis posits a window of opportunity exists to challenge unequal gender relations post-disaster. It suggests that as women may take on "nontraditional" gender roles within reconstruction activities, or be forced to enter paid employment post-event, this can change how they see themselves and how they are seen by men and the wider community, and this may improve unequal gender relations. The post-Mitch narrative I helped to construct questioned this thesis. My research constructs Nicaragua as a highly patriarchal country, with gendered inequalities of power clear within households and communities. Patriarchy has a varied and contested history of usage among social scientists (see Beechey, 1979). While most usually it is understood as male domination over women, some definitions also highlight the importance of male hierarchical social relations that establish an interdependence and solidarity between men, which in turn allows the domination of men over women. Rather than imply that every man is in a relatively dominant position, and every woman in an oppressed one, this approach allows for a more nuanced understanding of men's situations as socially constructed and lived relative to other men, as well as a focus on the impact this has on women.

How a woman experiences patriarchal control varies over time and space and indeed over her own life course. As Walby's (1990) work in the UK highlighted, the form patriarchy takes can change and be changed, but does not necessarily lessen. For example, women's entry into the labor force may not reduce patriarchy but, rather, reflect a shift in sites from private to public patriarchy, as women move from the control of their fathers, brothers and husbands in the home to the control of their male bosses in the workplace. Such shifts in how patriarchy is experienced by both men and women may be brought about by crisis, not just economic crisis forcing women into the work place out of necessity, but other forms of crises, such as those provoked by extreme climatic events and their aftermath. The nature of post-disaster relief and reconstruction may see individual patriarchal control lessened but an institutional patriarchal control taking its place. It has been suggested that the very nature of the asymmetrical power relationship within which humanitarian assistance is bestowed from First to Third World, and from male to female recipients "symbolically disempowers" women (Hyndman and de Alwis, 2003, p. 218), but what of men?

This chapter returns to the research I undertook post-Mitch and places men as central to the analysis to explore how patriarchal relations might be affected by an extreme climatic event and what this might mean for men and male identity post-disaster.

The research context

Living in Nicaragua and working with the feminist NGO Puntos de Encuentro at the time of the hurricane led to the opportunity to design and implement an in-depth study in four low-income communities. The study sought to understand how women and men experienced the hurricane and the relief and reconstruction process, to examine women's and men's perceptions of gender roles and relations and processes of decision-making, and to explore changes, if any, brought by Mitch. The study utilized a detailed questionnaire. In every household in the communities studied, the female partner of the male head or, when no man was present, the female head of household was approached. The questionnaire was followed by a semistructured interview with those women who agreed to this. A sample of the women's male partners were also surveyed and interviewed. The qualitative interviews were intended to contextualize and better understand the quantitative findings, and as such themes did not so much emerge from the qualitative analysis as were imposed. The women's interviews were analyzed first, effectively setting the narrative, with men's voices used to elaborate around these themes. Through various publications based on these findings, it became "known" how women and men experienced Mitch and this became part of the gendered knowledge base of disasters in the developing world (Bradshaw, 2001a, 2001b, 2002a, 2004, 2013; Bradshaw and Arenas, 2004; Bradshaw *et al.*, 2002).

A recent review of funded research projects found gender analysis often means that men (if present at all) are presented as a counterpoint to women in a compare-or-contrast framework (Bradshaw *et al.*, 2015). This is a charge that could also be leveled at many of my existing post-Mitch publications as men's voices in these are often confined to quotations used to introduce new sections, placing them outside the main analysis and effectively "othering" them. They are otherwise positioned as contradicting women's voices or as those seeking to limit women's abilities to speak and decide. This is not unusual. In the late 1990s, Cornwall (1998, p. 46) noted that gender roles and relations had become shorthand for inherently oppositional relations constructing men as "lurking" in the background and imagined as "powerful and oppositional figures." Conversely, and simultaneously in the gendered development discourse, men had been constructed as lazy—"sitting round talking" while women work (Whitehead, 2000). This image of idle men and hardworking women is also present in contemporary post-disaster images and reports (Mishra, 2009).

So, what would happen if men were no longer left "lurking" in the background but their voices were to be foregrounded? What new findings would emerge if the interviews were not interrogated for evidence to support the quantitative findings or as a contrast to women's opinions, but analyzed as stand-alone male narratives? Revisiting 15 of the qualitative interviews undertaken with men post-Mitch, this chapter explores what placing men as the subject of the analysis tells us about how patriarchal relations are reconstructed post-disaster.

The current gendered narrative of Hurricane Mitch

The study's existing narrative around the gendered impacts of Hurricane Mitch is largely based on quantitative data methods (see Bradshaw and Linneker, 2016). In the qualitative discourse, men's voices were privileged only in accounts of women's rescue and relief actions during and immediately after the hurricane. In the relief period, men recognized women's role, but only when they were seen as doing something out of the ordinary or taking on actions usually considered male, and even then it was conceptualized as helping men. A sexual division remained; for example, one man described how men engaged in road mending "went ahead" with the women "coming behind." The study highlighted that women had also been active in the reconstruction process, often targeted by international nongovernmental organizations as the beneficiaries of resources—yet many did not feel they had benefited personally from this. The number of women in income-generating activities fell after the hurricane and it was not clear if this allowed their participation in reconstruction or if this was due to their participation. Whatever the case, while women valued their reconstruction work as making a contribution to the household, men had a conflicting view, and post-event they placed even greater relative value on women's paid employment. The original narrative highlighted that there was little evidence to suggest any changes in gender roles that Mitch brought about led to changes in how women's work was valued, by men at least, and thus little change in gender relations. It found no evidence to support the often taken-for-granted idea that intimate partner violence increased post-Mitch (see also Bradshaw and Fordham, 2013, 2014).

Overall, the picture to emerge was one of men as largely unchanging in their entrenched behavior and attitudes, despite changes in women's roles and women's perceptions of their contribution to the household. Given the lack of changes in male attitudes, it is useful to first examine preexisting gender roles and relations before reflecting on this reanalysis of the study findings.

Meet the men in the study

Most of the men in the study were married or living with someone before the age of 20. When they talked about marriage, most used words like "responsibility" and "obligation" or even saw marriage as a "great weight to bear." As Juan explains, when a man is married "he no longer feels happy standing by the side of his father." As part of their new roles and responsibilities, married men felt a pressure to set up home as well as to act in a certain way, including feeling they should "abandon vices" and stop "going around drunk." Many of the interviews highlight, however, that it is not so much marriage but having a child that brings a change for men; commitment to a woman comes not so much because she is the partner of the man but because she is the mother of his child. While some men talked of falling in love, generally the discourse was more practical than romantic, and for some men women partners simply take the place of their mothers. As Santos explained, he looked for a wife when he was 16 after his mother died and he "needed someone to replace her." Luis, when asked what he would do if his wife left him, stated: "I would look to find myself another one, to finish off bringing up the kids."

There is a clear construction of gender roles and what men and women should do, with women in the *casa* (home) and men in the *campo* (fields), and this is a largely unchallenged discourse based on the nature of men and women, often seen as designated by God. Yet the value of women's work is recognized, at least by some men. Jose highlights that if there is no woman to make the food then men cannot work; but, he appreciates this as more integral to his life, noting that without the support of a woman not only could he not work but "could not think of the future, nothing." However, while women are valued, they are valued for what they do and can do, with the implicit understanding that there are things they cannot and should not do—use a machete, for example, as they lack the strength of men.

As the division of labor is tied up with the nature of men and women and the perceived natural differences, when women do work in the fields it is seen as sad for them and their families, and as leading to "talk." Gossip is focused on women who cross the gendered boundaries of what is seen to be the norm or natural for men and women, and women being outside the home leaves them open to talk. The men interviewed noted their wives would tell them they wanted to go somewhere, but only a few suggested the woman would ask, or need, their permission to leave on a day-to-day basis. However, many women never leave home unaccompanied by children, sisters or sisters-in-law, since if a woman goes out alone she is constructed as being *en la calle* (in the street) and assumed to be up to no good. Thus for some men, such as Silvio, it is important to keep this from happening. He tells how he has "worked until exhausted" to ensure his wife did not have to go out of the home to work for an income. For Sergio, it is important his wife stays at home "working obediently" for if he allowed his partner go out on her own (what he conceptualized as "from house to house") then he would "lose control, respect, the criteria to be a man."

Men were asked if a woman could take the lead in the home, or be the head of household, and in general the response was a clear no, justified by a number of men by reference to the Bible. The suggestion was that woman was created not just secondly, but as second and subordinate to men. As head of household, a responsible man should provide for his children, and the mother of his children. Male irresponsibility was conceptualized as involvement in activities such as drinking and gambling. This is something noted in the gendered literature also, especially in terms of men withholding income for personal use and leaving women and children in what has been termed secondary poverty (Chant, 2003; Bradshaw, 2002b). For the men, women's responsibility was linked to obedience and this included a woman having food on the table when the man arrived home and his clothes ready for him in the morning. A number of men conceptualized a good wife as one who not only did these things but also understood her husband, respected his opinion and did not question his decisions. In particular, a good woman knew not to question a man when he had been drinking. Thus while drinking was seen by men as bad behavior for men, "picking a fight" while a man was drunk was seen by men as bad behavior for women.

Listening to the men's voices, a strongly patriarchal discourse still emerges which constructs gender roles as based on biological difference or divine intent and is used to justify unequal gendered power relations. As such, men suggested it was not "being *machista*" to have what many would consider to be highly sexist views and very traditional opinions on women's roles, for male authority was seen to be the natural order or the "word of God."

The impact of Hurricane Mitch on masculine identities

Rather unusually, it was suggested that more men than women died during Mitch (Gomáriz, 1999) and my original explanation of this highlighted male *machismo* and resultant risk-taking in the face of danger. However, in revisiting the interviews, notions of responsibility and obligation are clear in the male narratives —and it is this that may have kept them in the path of danger as they tried to protect homes and goods. For women, while gendered norms suggest it is best they stay in the home, they are not confined to the home. Thus, women may have readily left for higher ground and safety with or without discussion with men.

As noted in the existing post-Mitch narrative, women and men were both actively involved in helping others during the event and in the immediate aftermath and Mitch seems to have been an eye-opener for some men. Juan notes that watching neighbors fleeing their homes while trying to salvage what they could highlighted to him the "utility" of women and the value of having a female partner to help. However, it was not women's actions alone that promoted male recognition of their utility, but what was important was how other, influential, men such as leaders of community groups and civil defense units evaluated women's actions. Jose, for example, highlighted that women's actions around health were useful but needed "order to be brought to them" by (male) leaders. Thus even as private patriarchal control by male partners may decline, so it is reimposed here by other, more public male actors.

Santos noted that Mitch made visible that women have a role to play in "*el campo y la casa*" (the fields and the home), but more generally work in the *campo* was still seen to be too heavy for women. While women did undertake such work post-Mitch, Nestor suggested that women's participation in this case was "illogical" if there was a man in the home as he should be the one to do this. For widows, there was no choice: Roger noted they had to "work like men" after the event. Yet many women, even those with a male partner present, stepped up to take on heavy work such as that involved in post-disaster food-for-work schemes. What was stopping men from undertaking this work? When men were asked what they did post-Mitch, the usual reply, even when they now had no land to farm, was "farmer," highlighting how closely men's identity is linked to their work. When asked the same, women would say "nothing" or "housework." Being a farmer not only constructed men as working, and thus having no spare time, but also made rehabilitation of the land a priority for them. When Cristiano talks of the "sadness" Mitch provoked and that "there is no happiness anymore," it is in relation to losses, not of life or home, but of the land.

As women were not seen to be farmers and did not self-identify as working, given the lack of paid work available at the time, they could and did take part in reconstruction projects. Cristiano suggested women's participation in projects in the circumstances was not a problem as it allowed them to have "some fun" and was a way of giving them a "little liberty." The discourse presented by some men did suggest that women's work in reconstruction meant more than "fun," addressing not just women's so-called practical gender needs but also their strategic gender interests (Molyneux, 1985; Moser, 1989). In terms of the practical need for women to provide food or water for the family, for example, Carmelo notes a "100% change" for women post-Mitch as they now "work early in the kitchen and then go to work the land." As argued in earlier publications, such changes not only escalate the hours women work but also reinforce existing gender roles by merely allowing women to better perform their existing role as "good" mothers.

Even men with strong traditional views accepted such work for women because it was seen to be a necessity, a product of the crisis. However, some men saw these new roles as advancing women's more strategic gender interests also, such as women gaining voice or promoting gendered rights. Jose suggests that because of changes brought by Mitch "women learnt to speak" and "even about agriculture." However, women's voices were still bounded in terms of how they were heard and by whom. Luis, while recognizing that women at times have "more ideas than men," notes that still men won't listen to them—for, as discussed above, women can't lead men.

Any shifts in gender relations from hurricane-related changes in gender roles were bounded by traditional ideas which remained, or became more ingrained, due to the event and post-Mitch impacts on male identity. When asked if the relations between men and women changed post-Mitch, Cristiano notes "what the hurricane changed was the land" not relations between men and women, while Jose notes "what has changed is the lack of work, but the love, it is still there." Indeed, Pedro thinks Mitch brought couples closer together as now, given the lack of resources, they "have to help each other and get along." However, rather than a new shared responsibility, most men seemed to feel pressure to provide and to be seen to provide. Carmelo notes "nights when I cannot sleep for thinking about how I am going to get food for the family without money."

When asked about other periods in their lives when they had experienced economic crisis, men recognized that at times the solution did lie with women, who would cook food to sell, for example. However, for Luis, if a woman was working post-Mitch and not the man, then people would talk. Pedro sees that such a man is no longer the head but is "irresponsible, not looking hard enough to find a way to provide for the household." This adds to a demasculinizing discourse which suggests that, when a woman earns instead of the man, she "acts as if she is the man," leading Nestor to conclude that the man should feel ashamed if the "woman is bringing the food" and he is just "sat at home" like a woman. For many men, while recognizing women's contribution to household resources as a help, it was very difficult to accept this help since it challenged their identity as the provider and their understandings of what makes a responsible man. This is interesting as the notion of irresponsible men is generally linked to those who spend what little money they have on drink, gambling or other women, hence keeping the household in secondary poverty. Here, however, men are being constructed as (or constructing others as) irresponsible, not through "bad" behavior but through inability to be "good" men.

In times of economic crisis, men reported they would usually turn to friends; but, after Mitch, all friends had also suffered losses so these male–male social networks did not work in the same way. In pre-Mitch times of crisis, men talked of how they felt "demoralized" by having to ask friends for help. In times of generalized crisis it might be assumed this feeling would be magnified, especially if such a request were rejected. That being said, some positive changes were reported in male–male relations. Some men spoke of their new faith in God which led to fewer alcohol-fueled conflicts and less male–male violence. In contrast, violence against women remained constructed by men as largely due to women's "jealousy" and women's inability to refrain from answering back to a drunk man. In this male discourse around disaster-related violence, men called again on ideas of responsibility. For example, Roger noted that when he had a few drinks after Mitch because he felt "weak, battered" it prompted criticism at home. He believes when a man drinks women assume they need "rescuing" and so respond by "scolding him or advising him." He defends his choice to drink post-Mitch, his irresponsible behavior, through a discourse of responsibility, noting he drank "because we're responsible for the business, the children" but women imagine that "just by having a drink that man's a goner, he's not worth anything anymore."

Revisiting the men's interviews suggests that, as is the case for women, how resources are generated matters for men. In this case, it is in terms of men's self-identity and also how they are viewed by other men, feeling other men would talk about them as a consequence of them not fulfilling their male roles. Men reported feeling sad and "afflicted" by the lack of work and their lack of ability to provide for the household. For Silvio, men who can't provide have "lost control." Interestingly, he included among this group those men involved in reconstruction projects such as self-build housing. The original narrative highlighted the high costs to women of participating in reconstruction activities; however, it seems that participation in post-disaster reconstruction may, at least for some men, also come at high cost.

The discourse highlights that what other men think matters to men. Constructing others as not being responsible men helps reestablish their own self-identity as the male head. Moreover, other men construct those who cannot provide for their families not as hurricane victims but as irresponsible in their inability to respond as men should to the crisis. Such harsh constructions might be best understood within a context where all men suffered losses, including to their identity as workers, providers and protectors.

Conclusion

Revisiting my original post-Mitch publications, it is clear men were "lurking in the background" of the analysis, constructed largely as conflictive, confrontational and controlling, with the communities constructed as highly patriarchal and marked by clear gender inequalities of power within couples. The reanalysis of the men's interviews in isolation from those of their female partners aimed to allow male voices to be more clearly heard to see if they would trouble this established narrative.

Applying a lens of masculinity did allow male voices to be heard and their gendered experience of disaster to be better understood. However, overall the new narrative presented here remains largely the same as the old. How gender roles and relations were understood by the men conformed to the original analysis, with men seeing their dominant role as natural or God-given and their role as provider a heavy responsibility uniquely falling to men. The discourse post-Mitch did see some shifts in how gender roles were conceptualized, with the crisis "allowing" women to share the responsibility to provide for the home. However, gender relations, for example, that dictate that the man should be the head of the household, remained unchanged. In part, the challenges that changing gender roles potentially posed to male dominance were addressed through a restatement of what being a responsible man means. The idea of how a responsible man acts shifted, moving away from seeing irresponsibility as associated with activities such as drinking, and toward understanding men's inability to act well, even if directly related to the disaster, as "irresponsible." These understandings were constructed through male–male relationships, as men sought to reestablish their masculinity in the new terrain, questioning the responsibility and thus the masculinity of other men to better reassert their own. This in turn allowed a reconstruction of the strong male–female patriarchal relations found before the event. While the nature of how patriarchal control was exercised might have shifted, it did not necessarily lessen for women, suggesting that patriarchal relations are highly resilient to disasters.

64   *Sarah Bradshaw*

Note

1 Many thanks to all those who supported the original research in Nicaragua: Puntos de Encuentro, in particular Tere Hernández, CIIR/Progressio, in particular Patricio Cranshaw, and to Oxfam GB for funding the original study. Many thanks again to the research team and to the men and women who took part in the study. Thanks to Brian Linneker for reading earlier versions of this paper.

References

Beechey, V. (1979) On patriarchy. *Feminist Review*, 3, pp. 66–82.
Bradshaw, S. (2001a) *Dangerous liaisons: Women, men and Hurricane Mitch*. Managua, Nicaragua: Puntos de Encuentro. Available from https://drive.google.com/file/d/0BzsYq-aDZJuZNTA4NmFmYjktOGNiZi00YjBmLTlmZDMtZmM1NzMyYzA1NTg5/view?pli=1 (accessed November 9, 2015).
Bradshaw, S. (2001b) Reconstructing roles and relations: Women's participation in reconstruction in post-Mitch Nicaragua. *Gender and Development*, 9(3), pp. 79–87.
Bradshaw, S. (2002a) Exploring the gender dimensions of reconstruction processes post-hurricane Mitch. *Journal of International Development*, 14, pp. 871–879.
Bradshaw, S. (2002b) *Gendered poverties and power relations: Looking inside communities and households*. Managua, Nicaragua: Puntos de Encuentro. Available from https://eprints.mdx.ac.uk/4031/1/genderedpoverties.pdf (accessed November 9, 2015).
Bradshaw, S. (2004) *Socio-economic impacts of natural disasters: A gender analysis*. United Nations Economic Commission for Latin America and the Caribbean (ECLAC) Serie Manuales 32. Available from http://www.cepal.org/en/publications/5596-socio-economic-impacts-natural-disasters-gender-analysis (accessed November 9, 2015).
Bradshaw, S. (2013) *Gender, development and disasters*. Cheltenham/Northampton, MA: Edward Elgar.
Bradshaw, S. and Arenas, A. (2004) *Análisis de género en la evaluación de los efectos socioeconómicos de los desastres naturales*. Comisión Económica para América Latina y el Caribe (CEPAL), Serie Manuales 33. Available from http://www.cepal.org/es/publicaciones/5597-analisis-de-genero-en-la-evaluacion-de-los-efectos-socioeconomicos-de-los (accessed December 9, 2015).
Bradshaw, S. and Fordham, M. (2013) *Women and girls in disasters. Report*, Department for International Development, UK, August. Available from http://www.gov.uk/government/uploads/system/uploads/attachment_data/file/236656/women-girls-disasters.pdf (accessed November 9, 2015).
Bradshaw, S. and Fordham M. (2014) Double disaster: Disaster risk through a gender lens. In: Collins, A. (ed.) *Natural hazards, risks and disasters in society*. Oxford: Elsevier, pp. 233–256.
Bradshaw, S. and Linneker, B. (2016) The gendered destruction and reconstruction of assets and the transformative potential of "disasters." In: Moser, C. (ed.) *Gender, asset accumulation and just cities: Pathways to transformation*. Oxon: Routledge, pp. 164–180.
Bradshaw, S., Linneker, B., Nussey, C. and Sanders-McDonagh E. (2015) *Evidence synthesis research award of the ESRC-DFID joint fund for poverty alleviation research: Gender theme*. Report. ESRC/DFID, UK, September. Available from http://www.theimpactinitiative.net (accessed March 21, 2016).
Bradshaw S., Linneker, B. and Zúniga, R.E. (2002) Social roles and spatial relations of NGOs and civil society: Participation and effectiveness post Hurricane Mitch. In:

McIlwaine, C. and Willis, K. (eds.) *Challenges and change in Middle America: Perspectives on development in Mexico, Central America and the Caribbean*. Harlow: Prentice Hall/Pearson, pp. 243–269.

Chant, S. (2003) *Female household headship and the feminisation of poverty: Facts, fictions and forward strategies*. London: London School of Economics and Political Science, Gender Institute. Available from http://eprints.lse.ac.uk/574/ (accessed November 9, 2015).

CIETinternational (1999) Principales resultados de la auditoría social para la emergencia y la reconstrucción—fase 2. Managua, Nicaragua: Civil Coordinator for Emergency and Reconstruction.

CIETinternational (2001) Principales resultados de la auditoría social sobre la condición de la pobreza—fase 3. Managua, Nicaragua: Civil Coordinator for Emergency and Reconstruction.

Cornwall, A. (1998) Gender, participation and the politics of difference. In: Guijt, I. and Shah M.K. (eds.) *The myth of community: Gender issues in participatory development*. London: Intermediate Technology Publications, pp. 46–57.

Cupples, J. (2007) Gender and Hurricane Mitch: Reconstructing subjectivities after disaster. *Disasters*, 31(2), pp. 155–175.

Delaney, P. and Shrader, E. (2000) *Gender and post-disaster reconstruction: The case of Hurricane Mitch in Honduras and Nicaragua*. Decision review draft. World Bank, January.

Gomáriz Moraga, E. (1999) Género y desastres: Introducción conceptual y análisis de situación. Paper prepared for the IDB technical meeting on the effects of Hurricane Mitch on women and their participation in the reconstruction of Central America.

Hyndman, J. and de Alwis, M. (2003) Beyond gender: Towards a feminist analysis of humanitarianism and development in Sri Lanka. *Women's Studies Quarterly*, 31(3–4), pp. 212–226.

Mishra, P. (2009) Let's share the stage: Involving men in gender equality and disaster risk reduction. In: Enarson, E. and Chakrabarti, D. (eds.) *Women, gender and disaster: Global issues and initiatives*. New Delhi: Sage, pp. 29–39.

Molyneux, M. (1985) Mobilization without emancipation? Women's interests, the state, and revolution in Nicaragua. *Feminist Studies*,11(2), pp. 227–254.

Moser, C. (1989) Gender planning in the Third World: Meeting practical and strategic gender needs. *World Development*, 17(11), pp. 1799–1825.

Walby, S. (1990) *Theorising patriarchy*. London: Wiley-Blackwell.

Whitehead, A. (2000) Continuities and discontinuities in political constructions of the working man in rural Sub-Saharan Africa: The "lazy man" in African agriculture. *The European Journal of Development Research*, 12, pp. 23–52.

Part II

The high cost of disaster for men

Coping with loss and change

6 Men, masculinities and wildfire

Embodied resistance and rupture

Christine Eriksen and Gordon Waitt

This chapter investigates narratives about men who manage wildfire in Australia. It builds on the growing body of work that presents firefighting as a gendered and spatially differentiated activity in Australia and North America (Enarson, 1984; Childs *et al.*, 2004; Yarnal *et al.*, 2004; Childs, 2006; Desmond, 2007; Maleta, 2009; Pacholok; 2009, 2013; Eriksen, 2014). This work argues that the privileged subject of the wildland firefighter is cast by discourses of (predominantly white) masculinities that position the bodies of men on the frontlines of fire as heroic, capable, physically strong and rational. The everyday narrative and performance of a place-based firefighting masculinity are so ingrained that they result in conscious and subconscious avoidance of appearances or allegations that align bodies with dominant understandings of femininity. Hence, the workplace and subject of the wildland firefighter is seemingly stabilized through the performance of a firefighting masculinity that includes the display of a masculine swagger, crude language and sharing stories of firefighting or heterosexual conquests.

The performance of a place-based firefighting masculinity trades on ageism, sexism and homophobia that dispute the worth of the bodies of women and other types of male firefighters (e.g. see Pacholok [2009, 2013] on contesting masculinities between structural and wildland firefighters). Such gendered practices are the focus of affirmative action and broader workplace policy concerns around the role of women and ethnic diversity within wildland firefighting and disaster management. The competencies of people who do not conform to the dominant gender norms are hidden behind bravado and cultural expectations that favor white, heterosexual men.

We employ a narrative approach to explore "being" and "becoming" a man within the context of wildland firefighting for the New South Wales (NSW) National Park and Wildlife Service (NPWS). The theoretical lens threads the "spatial imperative of subjectivity" (Probyn, 2003, p. 290), with performativity (Butler, 1993) and the notion of hegemonic masculinity (Connell, 2005). This theoretical approach enables us to think of how dominant discursive "regulatory fictions" of a wildland firefighting masculinity are either ruptured or made resilient through a repetitive set of acts and sayings that are always spatially situated (Butler, 2010). The chapter therefore takes a step back from thinking of gender as fixed and prescribed by biology. Instead, bodies become gendered through "doings," including sayings, and are therefore open to change. We conceive of fire management as a medium through which people get to know themselves and build relationships with others, rather than working within preconfigured social categories of gender that are expressed through firefighting. This requires us to think critically about how fire management may bring into existence particular gendered ways of being and becoming in particular contexts.

We argue that important insights of "being" and "becoming" a man are offered by exploring the contingencies of the material, social and cultural forces that form firefighting bodies and spaces as "events." In examining how gender emerges as particular events, we argue that the doing and undoing of the gendered lives of firefighters involves all persons in negotiating the hegemonic discursive regulator fictions of a firefighting masculinity. We present insights from men and women who manage wildfire to explore the following two related questions: How do individuals whose paid job description includes managing wildfire become men? How are individuals rupturing dominant norms of "being" and "becoming" a man who manages wildfire?

Methodology

We examine issues and themes about men and masculinities in wildland firefighting raised by NSW NPWS employees in semistructured interviews. Twenty-seven interviews were conducted during July–August 2011 and August 2013, with nineteen women and eight men. All were involved in wildland firefighting but in different capacities as rangers, field officers, planners, logistical personnel or in communications. Interview participants volunteered to take part after invitations were extended via email through the NSW NPWS Head Office. An interview schedule was designed to explore participants' gendered sense of self through their workplace-based practices and conversations. In order to explore changes in the gender workplace politics and policies, the interviews were structured around two themes: 1) why participants chose a career in wildland firefighting; and 2) how participants negotiate everyday gender relations, traditions and identities. Interviews were conducted as open conversation rather than structured questions and answers, to explore in-depth topics of concern to participants and themes that emerged during the discussion.

The female lead author, an academic with national and international wildfire research experience, a rural upbringing and basic firefighting training, conducted the interviews. All these attributes may have influenced or shaped the particular stories shared by participants depending on perceptions of common ground, points of disagreement or established trust. Interviews occurred at a location of the participants' choosing where they could not be overheard or interrupted to ease any potential discomfort or concern relating to discussing workplace issues or emotionally charged stories. They lasted between 45 and 90 minutes and were audio recorded and transcribed verbatim. Using the Computer Assisted Qualitative Data Analysis Software (CAQDAS) NVivo v.10, the transcripts were systematically coded using both a priori themes, such as training and task delegation, and emerging themes, such as self-confidence and sexism (Bazeley, 2007; Riessman, 2008). The direct interview quotations illustrate how the workplace is a crucial site for reinforcing or challenging ideas about men who fight fires and wildland firefighting masculinity.

Men who manage wildfire: doing firefighting masculinity

At first thought, many research participants expressed a belief that discrimination along the line of gender is an issue of the past, overcome by a generational shift within the NPWS, and society more broadly (Johnson, 2012). Among research participants, there was a greater awareness of gender privilege within firefighting, and a clear sense of the importance of facilitating gender equity. For example, one long-serving man who manages wildfire explained that:

> I think the norm is that firefighting is a boy's role, and I think we're in a time of change. I believe that there's a huge role for women to play at all levels of firefighting. I believe it's just about changing attitudes ... I think in Parks [sic] there's less of that machismo and more of that acceptance and nurturing process [than in other organizations]. I'm not saying "nurturing" in a condescending way, I'm just saying that there is that process. I think that's really important.

> (Male Ranger, Aug. 2013)

Rangers were aware of challenging the gender privilege given to men who manage fire within the institutional reproduction of firefighting as a "boy's role." Participants were alert to how the work of field officers (mainly physically demanding work outdoors) is embedded in the performance of a firefighting masculinity, including biological reductionism that positions women as both physically weaker and more emotional than men. Upon greater reflection, during the interviews, the question of how to address the inherent "machismo" embedded within wildland fire management culture was far from straightforward. In part, this is because troubling gender inequity worldwide during the past five decades is ongoing, meaning "prejudice and sexism remain firmly embedded in social structures" albeit often latent or disguised in equal opportunity policies (Fordham, 2004, p. 178). The wildland firefighting profession is no exception (Enarson, 1984; Childs, 2006; Desmond, 2007; Pacholok, 2013; Eriksen, 2014).

Participants spoke of a lingering culture of gendered and classed discrimination around certain positions, particularly the roles of field officer and aviation crew. The aviation branch was consistently described as a "boys' club." This view is consistent with statistics: 91% of trained NSW NPWS Aviation Specialists in 2013 were men (n=89). The following aviation pilot narrative is illustrative:

[Interviewer: You mentioned the stone wall that she (female aviation crew member) faced when she first arrived. How does that play out in the everyday work environment?]

I think observationally it plays out in that I don't think she's given the same respect as the other people of the same skill set or experience are given. It's a harder road. She has to prove herself. When I think about how there was a male and a female start at the same time. Watching their supervisor put them through the paces and do the training, the treatment to her was more robust, let's say, and harsher, to be honest. So she had to do many more briefings and practices before she was allowed to do it for real than her male counterpart.

[Interviewer: Without any apparent reason?]

Oh, in fact, it panned out that she is a much better operator than her male counterpart, who was let go early and was found wanting. So he's got back under supervision.

[Interviewer: It's kind of ironic, isn't it? The fact she was grilled longer would mean that she was more prepared once she was deployed.] It is. Yes, I think you're right. Had the male counterpart gone under the same regime, I think that he would have been all right.

(Male Project Manager, Aug. 2013)

This narrative illustrates how a firefighter who assumes that men are better positioned to undertake the work of aviation crews (or any other firefighting role) than women are taking the position of hegemonic firefighting masculinity. Connell (2005, p. 77) defines hegemonic masculinity as "the configuration of gender practice which embodies the currently accepted answer to the problem of the legitimacy of patriarchy, which guarantees (or is taken to guarantee) the dominant position of men, and the subordination of women." Hegemonic masculinity is not fixed, nor the same everywhere, but it occupies a dominant position within particular patterns of gendered relations.

The doing of hegemonic masculinity is again reflected in participants' accounts of the "tap on the shoulder," a practice described as the unofficial method used to single out staff to temporarily act in higher positions or be shortlisted for competitive positions, such as on helicopter crews. This practice of promotion is not just gender biased, but sustains the symbolic imagery of physically strong, heterosexual, outdoorsy men as *real* firefighters; it is also usually restricted to men who present themselves in this way. This "venerable rural myth of rugged individualism" (Campbell *et al.*, 2006, p. 2) also defines the standards of competence valued highest within the male-dominated world of firefighting (Desmond, 2007). Pacholok (2009) argues that male firefighters use gendered strategies of self to navigate hierarchy because hegemonic masculinity is always contested rather than statically reproduced (Connell, 2010):

If hegemonic masculinity is a given these tactics would not be required for those with the most power and status. However, because the positions at the top of the gender hierarchy are never secure, even those with power … are compelled to engage in practices that refute the integrity of those they perceive as Other.

(Pacholok, 2009, p. 494)

Some men benefit from hegemonic masculinity without doing anything. Connell (2010) refers to the benefits men get from the subordination of men and women who do not live up to the ideals, as patriarchal dividends. Equally, there are some men who are subordinated by hegemonic masculinity. As one female ranger explained:

> I've always felt it's an advantage to be a woman. It feels like you're playing it both ways somehow. If you're a male ranger dealing with field officers, there's certain expectations of them, which is probably as difficult to deal with as maybe the lack of expectation of the females. So in some ways as a female, I've always felt you have a bit more freedom. You can kind of jolly them along and yet still get to do what you want to do. I don't know how that works for the male rangers, whether they really have to fit in with the guys and behave a certain way.
>
> (Female Ranger, July 2011)

According to this female ranger, it is men who have to "fit in" with the hegemonic gender norms. Some men who manage wildfires can thereby be as constrained by hegemonic masculinity as women, as they navigate the socially constructed, historically situated, and limited expression of a firefighting masculinity. Women who manage wildfires, although disadvantaged by hegemonic gender norms, may accrue certain advantages from reproducing a "complicit firefighting femininity." Some women may be complicit by tapping into essential ideas of femininity, such as accepting overt sexist language, jollying men along, choosing an offer of help over carrying a heavy piece of equipment or accepting the notion of being one of the "girls." At the same time, the experience of women who manage wildfire as rangers suggests being one of the "girls" often plays out in terms of disempowerment through gendered job allocation:

> At the end of the day, they [male colleagues] said, "How was your day?" I said, "That was atrocious guys! Why did you throw me in that [negotiating with a local farmer]?" They said, "Oh we did it on purpose because you're a female and we know that girls can talk better to farmers," like 'cause this guy had been in trouble with National Parks 'cause he kills wildlife. I said, "Next time I want you to brief me on this so I know what I'm going into! Second, you throwing me at that, that's not a very nice thing to do!" "Oh yeah, but girls, you're better at negotiating," which is probably true and cocky farmers tend to calm down a bit 'cause it's like, "I've got to speak to 'the girl'" but it's still, you get given those roles and had I gone in briefed, I may have been a bit more prepared to deal with that aggression. So in that regard I suppose you do get picked out for those roles. I've just got used to it now being one of the only girls.
>
> (Female Ranger, Aug. 2013)

How a gender order is reproduced within the patriarchal structures of firefighting is linked to visible and invisible resistance to change. This includes becoming complicit with naturalized gendered roles and responsibilities that are aligned to a sexed body, as in the above notion that women are naturally better oral communicators.

Furthermore, there is a shared narrative that implies that for women to become wildland firefighters required a cohort of women to become complicit in reproducing gender norms that converge men and masculinity with management of wildfire. As one woman who manages wildfire explained:

> There was a group of women before me in National Parks who broke down a lot of the barriers for women in National Parks, not just in firefighting, but generally. They're very strong women. They basically had to beat the men at their own game. They had to be as tough as any other man and because they were the ones that came through and broke down all the barriers, it was a lot easier for my generation who came through after.
>
> (Female Ranger, Aug. 2011)

This narrative reflects the work of academics who argue that women's inclusion into the ranks of firefighting has usually been on the proviso that they meet the perceived nonemotional, no-nonsense, noncompromising masculine way of engaging with risk (Yarnal *et al.*, 2004; Desmond, 2007). As Cupples (2007, p. 169) explains, although women "are disadvantaged by the existence of essentialist cultural constructions, they also draw on these notions, sometimes strategically and often unconsciously, to make sense of their lives and to construct their subjectivities in ways that are discursively present" (and see Pacholok, 2009). Women who take up the seemingly masculine traits or attributes of a firefighting masculinity are arguably complicit with hegemonic norms of masculinities and femininities.

Furthermore, resistance to undoing a naturalized gender order is illustrated by everyday expressions that are used to define what it means to be a wildland firefighter. An example is the gendered language consciously and subconsciously interwoven into the narratives of men who manage wildfire that converge firefighting, warfare, masculinity and men:

> For boys, it [firefighting] is a little bit more natural I think. I'm not quite sure why but, you know, fighting fires and running amok, there's the volunteering, and all that type of stuff for that type of activity is a little bit more of a "huk, huk, huk"; it's sort of like being in the army but not being in the army, if you know what I mean, and I think women are more circumspect about that.
>
> (Male Ranger, Aug. 2013)

> I think there's still quite often this belief that firefighting is related to the military and that we're fighting a war. Personally, I find that a bit twee and it's nothing like it at all. But I think, you know, it's that bravado that kicks in.
>
> (Male Project Manager, Aug. 2013)

> I can't say I notice a [gendered] difference in approach to risk. There is unfortunately still a bit of a, you know, the male ego stuff with fires with some

people. The war story stuff and all that and you never get women doing that. I've never heard women tell war stories.

(Male Senior Field Supervisor, Aug. 2013)

Undoing gendered workplace tensions revolves around negotiating two cultural expectations of men who manage wildfire: 1) how the sexed body is aligned with a particular understanding of gender; and 2) the hierarchical patriarchal gender order that positions men above women in the process of decision-making. For example, one woman who manages wildfire outlined how "collegial acceptance" rests on her bodily proficiencies in performing a firefighting masculinity that involves skills such as the use of a rake hoe:

> I became more competent than them [male colleagues]. I would get in and muck in a lot more. I think that's partly my farming background. I remember a couple of the guys that I'd work with telling the other guys: "She's alright, as opposed to the rest of them." But it wasn't actually just about women. It was about rangers compared to field officers. It's almost like it's a blue-collar, white-collar tension. So when these guys would say, "She's alright," it was "she" as a woman but also "she" as a ranger. "Oh, she's got a degree but oh, she can still use a rake hoe."
>
> (Female Ranger, Aug. 2011)

In this example, acceptance as a ranger is based on reproducing the hegemonic gender norms through the physical training of bodies for strength demonstrated by using a rake hoe. The performance of gender among this group of wildland firefighters is revealed as being variously subliminal, engrained, conscious and complicit with hegemonic norms of a firefighting masculinity. It also demonstrates how men, gender and class intersect, as class difference becomes manifest in the enactment of the more bureaucratic, professional form of masculinity associated with rangers and senior management by field officers (and see Desmond, 2007; Pacholok, 2013).

Rupturing gender norms of men who manage wildfire

In contrast to the doing of the conventional gender order, hegemonic firefighting norms are also challenged, particularly by firefighters who embrace the principles of training and fitness programs emphasizing technique and stamina over physical strength. Both training and fitness programs provide evidence that the norms that narrow understandings of becoming a man who manages wildfire can be widened.

Training programs provide a context in which individuals spoke of becoming highly aware of gender differences, particularly through physical strength. One trainer positioned young men as an "unstoppable force," where women are positioned as "more circumspect and intelligent" problem solvers:

> When I'm mentoring somebody, I find with younger men they're the unstoppable force [laughs], and they're just waiting to run into that immovable

object, whereas women are more circumspect and intelligent about that. I'm not saying that women are not prepared to take the risk. They'll take the risk but it's just that the journey is more important than the destination. I think that is the fundamental difference.

(Male Ranger, Aug. 2013)

A seasonal firefighter drew attention to how the perception of men who manage wildfire as an "unstoppable force" can disempower all team members, as bravado may result in poor judgment:

If you have an individual who is not as confident in their competence, it's not reacted to differently if that's a man or woman. But I think a woman might say, "I don't feel confident doing that," which is good because everybody's aware of it and can make allowances for it. Whereas a guy, he might not want to think that he's not as good as everybody else, and that could put you all at risk if that's not properly recognized.

(Male Seasonal Firefighter, Aug. 2013)

Men who blatantly reproduce conventional firefighting masculinity can also be the cause of anger and discontent, as evident in the narrative of a senior field supervisor:

That macho thing is gradually disappearing. You still get the boys standing round leaning on the trucks, you know, laughing and carrying on, which I just avoid like the plague. I can't stand that stuff. We're here to do a job, just concentrate. If you've got nothing else to do look at the maps, find out what the weather is, that sort of stuff. I'm just tired of it. Years and years of it, not interested anymore. I think it is becoming less common, maybe because there are more women out on the fire line and stuff and, I mean, you can still have jokes and stuff but it's not that sort of chest puffing rubbish.

(Male Senior Field Supervisor, Aug. 2013)

Bravado or hyper-masculine firefighting behavior, while "gradually disappearing," narrows the understanding of what it means to be an individual who manages wildfire. Such behavior is frowned upon by many firefighters, for example, ridiculed as "chest puffing rubbish." As discussed by Eriksen (2014) and Desmond (2007), the bravado associated with a firefighting masculinity is both unnecessarily aggressive, even dangerously daring.

The above quotations illustrate the gendered dynamics of managing wildfire. Training to become individuals who manage wildfire becomes sites where shared gendered performances of a firefighting masculinity are reproduced and challenged. As one training coordinator explained, when the emphasis is shifted from physical strength to technical competence (as below) dominant gender norms no longer neatly align with biological sex.

I suppose because the base training that we do for firefighting involves working with pumps and being able to start pumps. They're [women] quite intimidated about how much effort it takes to start a pump. We now have electric start pumps and that's taken some of the fear away, but we still have to get them to be able to demonstrate that they can start it if the electric system's not working in an emergency. I've found that generally if you spend a little bit of time with them and show them the right techniques, they don't have any problems. Once they've done it once they're right, they fly. It's that little bit of confidence about having the strength. I think it's just a self-conscious thing that most females think, "I'm not physically as strong as what's going to be required." Once you show them the right techniques they realise that it's not about strength. It's about the actual way that you do it. It's the technique that you use rather than the strength that you've got.

(Male Training Coordinator, Aug. 2013)

Training techniques are challenging gendered understandings of what it means for individuals to become a wildland firefighter. Technique breaks narrow understandings of wildland firefighting that depend on the convergence of men, masculinity, bravado and physical strength.

Changes in fitness test requirements provide another example that illustrates how becoming a wildland firefighter is not aligned with physical strength. Instead, retaining the subject position of a wildland firefighter is achieved by what one participant called a "modified arduous" demonstration of the "fit" rather than "muscly" body. The demonstration of the fit body, as described below, is not understood as being aligned to the sexed and gendered body, and benefits both men and women:

Every year in order to be competent to participate in firefighting you have to undergo this fitness test. There are a number of levels: arduous, moderate or light—and those different levels are assigned to different roles in firefighting … The arduous test, you have to carry over 20 kilos on your back and walk 12 times round an oval, a standard athletics oval, within a certain amount of time. So if I wanted to do the arduous I was alongside men that were over 6 foot tall, big muscly field officers, and there were women smaller than me. For them, it was a really serious physical effort. People were getting in within a couple of seconds of the limit, particularly the shorter people. There are some guys that are in the same boat, who are shorter than me and about the same weight, and they were literally doubled over. In the last two years, they brought in a modified arduous for people that are 68 kilos and under. You do the same distance, the same time, but you only have to carry 15 kilos. So that was a concession that really supported women, but there are men that do the modified arduous as well. They weren't saying you have to be able to carry 20 kilos on your back to do remote bushfire work. What they're saying is you need to be this fit in order to do that. So the fact is that those people are that fit, they're just not as heavy and bulky as the bigger guys.

(Female Operations Coordinator, July 2011)

Training for skills rather than strength, and fitness rather than muscles, are important ways to highlight the competence and leadership abilities of both men and women firefighters whose bodies do not conform to gendered shapes and sizes of those who manage wildfire. Nevertheless, technique and stamina, as an alternative to strength, are invisible to the unknowing eye until observed and recognized as such.

Implications for change

Men and women recounting workplace narratives for the NSW National Park and Wildlife Service suggest that gender continues to be a key influence on how they experience wildland firefighting. Historical ideas framing gender within fire-fighting institutions, particularly with respect to outdoors work, still powerfully define acceptable and unacceptable gendered behavior, notwithstanding decades of affirmative action, and training techniques breaking the link between physical body strength and capability. Different gender expectations and experiences persist for men and for women. The presence of women on the fireline both challenges and reinscribes gendered concepts and practices of the bodies of individuals who manage fire.

For individuals to become men who manage wildfire the desire to gain the approval of peers may lead to performing a firefighting masculinity that disempowers women. These are men who may condone equal opportunities in the workplace but might not have questioned the ways in which their own behavior reproduces gendered inequalities and sexism. Masculine privilege, ironically, extends to men's assertion of their "protection" of women perceived as being "at risk," a dynamic in which women may be complicit by accepting this subordination. This protective behavior, perhaps well meant in a stereotypically chivalrous or gentlemanly way, is also strongly linked to the ideology of heterosexuality that configures male bodies as strong, rational, bounded and authoritative. The ongoing bravado demanded of those men who strive to maintain self-esteem, confidence and respect through "heroic" action on the "battlefield" of firefighting poses ongoing physical and mental health challenges for men (Pacholok, 2013). These patterns are consistent with arguments outlined by many feminist researchers on how the gendered and sexed dimensions of the body both express difference and sustain inequality and inequity (Grosz, 1994).

Current approaches to firefighting training that prioritize technique over physical strength alone will not trouble the cultural expectations that people uphold for wildland firefighters. To negotiate this complexity requires engaging individuals in education and mentoring programs that challenge the naturalized link between men, masculinity and the management of fire. One helpful starting point is programs that provide opportunities for individual and group reflection (Albury *et al.*, 2011). Reflection offers one key aspect of an educational and mentoring response toward preventing sexism and gender inequity. It may be used as part of training programs, first by reenacting a fire management "event," then moving from describing what happened to what participants were thinking and feeling, and then to evaluating what was "good" or "bad" about the experience. The next stage involves making sense of what happened and thinking through alternatives that help in designing an action plan. Finally, these deliberations offer an entry point to scale up from the individual to the social by locating how wildland firefighters' lives are different, situated within historical, cultural and gendered contexts. Through these discussions three challenges are posed to men who continue to embrace ideas that sustain conventional notions of a firefighting masculinity fashioned by bravado: first, how they reproduce and reify gendered power hierarchies that reaffirm women as the "weaker sex"; second, how they silence women's skills and accomplishments as firefighters; and third, how they narrow rather than widen understanding of men, masculinity and wildland firefighting.

References

Albury, K., Carmody, M., Evers, C. and Lumby, C. (2011) Playing by the rules: Researching, teaching and learning sexual ethics with young men in the Australian National Rugby League, *Sex Education*, 11(3), pp. 339–351.

Bazeley, P. (2007) *Qualitative data analysis with NVivo*. London: Sage.

Butler, J. (1993) *Bodies that matter: On the discursive limits of "sex."* New York: Routledge.

Butler, J. (2010) *Gender trouble: Feminism and the subversion of identity*. New York and London: Routledge.

Campbell, H., Bell, M.M. and Finney, M. (eds.) (2006) *Country boys: Masculinity and rural life*. University Park, PA: The Pennsylvania State University Press.

Childs, M. (2006) Counting women in the Australian fire services. *The Australian Journal of Emergency Management*, 21(2), pp. 29–34.

Childs, M., Morris, M. and Ingham, V. (2004) The rise and rise of clean, white-collar (fire-fighting) work. *Disaster Prevention and Management: An International Journal*, 13(5), pp. 409–414.

Connell, R.W. (2005) *Masculinities*. Cambridge: Polity Press.

Connell, R.W. (2010) *Short introductions: Gender*. Cambridge: Polity Press.

Cupples, J. (2007) Gender and Hurricane Mitch: Reconstructing subjectivities after disaster. *Disasters*, 31(2), pp. 155–175.

Desmond, M. (2007) *On the fireline: Living and dying with wildland firefighters*. Chicago: The University of Chicago Press.

Enarson, E.P. (1984) *Woods-working women: Sexual integration in the U.S. Forest Service*. Tuscaloosa, AL: The University of Alabama Press.

Eriksen, C. (2014) *Gender and wildfire: Landscapes of uncertainty*. New York and London: Routledge.

Fordham, M. (2004) Gendering vulnerability analysis: Towards a more nuanced approach. In: Bankoff, G., Ferks, G. and Hilhorst, D. (eds.) *Mapping vulnerability: Disasters, development and people*. London: Earthscan, pp. 174–182.

Grosz, E. (1994) *Volatile bodies: Towards a corporeal feminism*. Sydney: Allen & Unwin.

Johnson, L.C. (2012) Feminist geography 30 years on—they came, they saw but did they conquer? *Geographical Research*, 50(4), pp. 345–355.

Maleta, Y. (2009) Playing with fire: Gender at work and the Australian female cultural experience within rural fire fighting. *Journal of Sociology*, 45(3), pp. 291–306.

Pacholok, S. (2009) Gendered strategies of self: Navigating hierarchy and contesting masculinities. *Gender, Work & Organization*, 16(4), pp. 471–500.

Pacholok, S. (2013) *Into the fire: Disaster and the remaking of gender*. Toronto: University of Toronto Press.

Probyn, E. (2003) The spatial imperative of subjectivity. In: Anderson, K., Domosh, M., Pile, S. and Thrift, N. (eds.) *Handbook of cultural geography*. London: Sage Publications, pp. 290–299.

Riessman, C.K. (2008) *Narrative methods for the human sciences*. Thousand Oaks: Sage Publications.

Yarnal, C.M., Dowler, L. and Hutchinson, S. (2004) Don't let the bastards see you sweat: Masculinity, public and private space, and the volunteer firehouse. *Environment and Planning A*, 36(4), pp. 685–699.

7 Emotional and personal costs for men of the Black Saturday bushfires in Victoria, Australia

Debra Parkinson and Claire Zara

The likelihood of Australians facing natural disaster is high—estimated at a one in six lifetime exposure (McFarlane, 2005). The worst bushfires in Australia since settlement occurred on February 7, 2009 and are known as "Black Saturday." Classified as "catastrophic," these fires were devastating in their human impact, with 173 deaths, 414 people injured, 2,133 houses destroyed (Cameron *et al.*, 2009; Victorian Bushfires Royal Commission, 2010) and approximately 7,000 people displaced (Atkins, 2011). As very little gendered sociological research has been conducted with those directly affected by this or similar disasters in developed countries, this research with 32 men aimed specifically to document men's reflections on gender in this context. The men spoke at length of their terror, exhaustion and, often, of their powerlessness. Having survived, their recovery was plagued by the disaster's long-lasting impacts.

In Western society, as in most societies, men benefit from a gender hierarchy empowering men. However, some men benefit more than others, with intersections of class and race moderating privilege (Pease, 2010). Men's success in embodying and enacting versions of ideal manhood contribute to determining the level of privilege they enjoy. In the everyday world, the gender order is fragile, based as it is on a false dichotomy (Johnson and Repta, 2012). However, after disaster, any pretense of a "natural" fit between sex and gender enactment is exposed. The gender hierarchy depends on the public image of strong men in control and in charge and never more so than in an emergency. This has consequences for all, as noted by Victorian Assistant Police Commissioner, Tim Cartwright (2012):

> This is about men being men, as they see themselves, as we see ourselves, in response to disasters. The implications are that in public we are strong and fearless and not affected, but the implication for many women is when we come home, we don't cope at all.

This research followed our prior women-focused research, which found an increase in domestic violence following Black Saturday (Parkinson *et al.*, 2011; Parkinson and Zara, 2013; Parkinson, 2015). The chapter begins with the research design and a description of the sample. Following this, the men's narratives capture the loss of control they felt on the day of the Black Saturday bushfires and in its long aftermath. The men reflected on the weight of expectations of them—contrasted with their feelings of inadequacy—and the consequences for them in their homes and workplaces. Explanations of men's reluctance to seek help follow, and the subsequent emergence of harmful behaviors by some men are explored. The gendered vulnerabilities and risks that emerged for men through this disaster are discussed, with recommendations for change. An overview of one pathbreaking initiative, the Gender and Disaster Taskforce, concludes the chapter.

Methodology

The researchers were employed by Women's Health Goulburn North East with support from the National Disaster Resilience Grants Scheme. Employing a qualitative methodology, men were invited to speak frankly and in their own terms about their disaster experiences. Two researchers jointly conducted the interviews, which were digitally recorded with permission and transcribed verbatim. Informants received a copy of their transcribed interview and were invited to amend it at their discretion. Men aged over 18 who were fire-affected and in the shires of Mitchell and Murrindindi were invited to be interviewed. Theoretical sampling was used, and recruitment was through advertising in local media, flyers at community venues, and professional and community networks. The sample of 32 was self-selected, with some informants making contact with the researchers on their own initiative. Modified grounded theory guided analysis (Glaser and Strauss, 1967; Spradley, 1980) and coding was conducted using the software package NVivo v.10. Coding validity was enhanced by independent coding by both researchers, and informant checks. Ethics approval was received from both Monash University Human Research Ethics Committee (MUHREC) and North East Health HREC, and a Men's Advisory Group was formed to assist the researchers.

The 32 men profiled here were aged between 36 and 69 years, with a median of 14 years in residence. Five were single on Black Saturday; 24 were married or in a committed relationship; and three had more casual relationships. In the four years between Black Saturday and the interview, there were four separations and two of the three more casual relationships ended. Seven men were professional firefighters and five were Country Fire Authority (CFA) volunteer firefighters on Black Saturday. One was a State Emergency Service (SES) volunteer. Others had taken on these volunteer roles in past years.

Losing control on Black Saturday

For a week, temperatures had soared to over 40°C, each day bringing higher levels of threat in a tinder-dry landscape. The government policy of "Prepare, stay and defend or leave early" was tested to its limit. Power, landlines and cell phone connections cut in and out as the massive firestorm approached:

> It was a firestorm and some of the flames were blue. So it was a really high temperature … It's moving really, really quickly. It's a big, gaseous mixture

that lacks oxygen sucking air in and just goes "boom." It reignites, so really the sky's on fire. Just flames everywhere. It scared the hell out of me.

(Matthew, cited in Zara and Parkinson, 2013, p. 15)

In Australia, as elsewhere, the bravery and selflessness of firefighters is valorized, especially those in volunteer ranks who receive "near-universal admiration for [their] bravery and community spirit" (Manne, 2009, para. 12). Expected to be "manly" and "heroic" in disasters (Rivers, 1982; Scanlon, 1997; Elinder and Erixson, 2012), the 32 men interviewed pointed out the impossibility of living up to ideals of manhood in such a firestorm. In this context, "Given the expectations of traditional masculinity, trauma as a loss of control can be seen as a failure in masculinity and a failure to conform to one's self-conception" (Pease, 2014, p. 63).

Any semblance of control disappeared and gender conditioning fell away in the face of death. Lee spoke of his terror scrambling to the fire truck in conditions of zero visibility. The fire truck was in survival mode, whereby curtains on the truck were pulled down and the sprinklers activated. The driver was crying that no one was coming for them and they would die. Stuart spoke of one of his crew leaders who "lost the plot... just totally lost it," when having to drive through the fire front. The effect on many of the men was of being shaken to their core. Learnt behaviors of manliness were secondary to human reaction, as men spoke of publicly crying in response to such immense fear and tragedy. Big, burly, fire-fighting men were seen crying, "breaking out in tears," "bursting into tears" and "exploding into tears."

There is an intrinsic contradiction for members of top-down traditional "command and control" organizations to be so clearly out of control and at the mercy of the firestorm, especially as firefighters are the embodiment of hegemonic masculinity, and competence in the role is based on being calm in a crisis (Pacholok, 2009, 2013). Media and disaster management alike celebrate hegemonic masculinity (Cox, 1998; Enarson, 2006; Desmond, 2008), arguably because it is in their interests. Men who embody hegemonic masculinity, and sit in roles commensurate to that, are in charge of emergency management and in charge of the media that shapes our understanding of what happens in disaster.

Black Saturday tested men's ability to live up to the impossible. They were expected to measure up to the firestorm, to be brave, decisive, unemotional and stoic, and to not break down in its aftermath.

Men ... are supposed to have some sort of masculine, paternal role in managing any sort of situation ... where, when the fire starts, you just flick the switch and they become robotic. And I don't think you're meant to have feelings.

(Adam)

Loss of control in the aftermath

After surviving the fire, the men found that the world had changed—and a sense of place was lost for many (Proudley, 2008). Practical imperatives of housing and income coexisted with emotional turmoil for some. Much of the early initiative shown by women and men immediately after the fires dissipated over time. One informant told of "sensitive blokes" expressing their grief in the aftermath, as if such expression belonged only to a certain kind of man.

Regrets haunted survivors and, inevitably, relationships suffered. Several men spoke of trying to protect their wives or partners by hiding the effect the fires had on them. Others wanted to talk to their partners but felt constrained. These men questioned whether women are, in fact, intrinsically better at communication, quoting their wives' and partners' attitude as "Just get over it." As relationships deteriorated, they described merely "existing" side by side, and drifting apart and away. Paul said only three couples were still together from his friendship network of 16 couples.

Success in getting over the fires seemed to be measured and policed in the public domain. There were capricious proclamations of heroic status for some men, and for others judgments of failure, responses of pity and monitoring. Bernard recalled constant questioning: "Why haven't you got it together? Why haven't you got your garden fixed? Why haven't you got your house done yet? What are you doing with your life? Why haven't you gone back to work? Why haven't you?" And men did speak of throwing themselves into work or distractions. Some took on leadership roles within the community, expanding their position as head of the household into the public space. Others turned to traditional, masculine activities and took up tools to build or clear, or pens and laptops to apply for grants and permits.

For many men, this effectively blocked emotional responses and denied partners and families the support and intimacy they needed; and they denied it of themselves. Working for the family was, for some men, a way of avoiding being with the family, and all that might entail. Adam remembered his strategy: "Run away from the emotional stuff and try and get yourself bogged down into work." In response, women frequently had no choice but to turn their focus to the family, curtailing careers, leaving jobs or limiting hours of work. With men in the workforce and out in the world and women in the home, traditional gender roles were strengthened in this disaster's aftermath—as in so many other disasters. After the California wildfires in 1991, for example, Susanna Hoffman writes that "Progress in carving out new gender behavior suffered a fifty-year setback. In the shock of loss both men and women retreated into traditional cultural realms and personas" (1998, p. 57).

Workplace consequences

Men's refuge in workplaces was short-lived. With a clear deadline on sympathy, employing bodies often failed to offer accessible and personal debriefing or ongoing and confidential counseling. For men in Emergency Service Organizations (ESOs), there were career penalties for not coping. Men perceived to have let the trauma of the day and the tragedy of events "get to them" were relegated to a lesser status, judged as not quite measuring up to the hegemonic masculinity that characterizes emergency services.

ESOs and bureaucracies forged ahead on the flawed basis of men being men, robotic, in control and resistant to emotion. While women were asked to forgo their right to live free from violence in supporting men (Parkinson, 2015), no such sacrifice was expected of institutions. Two men told of the work demands they felt. One was asked to be involved in a job that would (in all probability) require him to be involved in the Victorian Bushfires Royal Commission and potentially relive the trauma and loss he was trying to overcome. His response was to say to his employer, "I'm not sure I can." As a man wanting to appear to be in control, Aaron did what he could to protect himself, but was nevertheless directed to undertake this work. In another example, Chris told of helping police and forensic teams locate and identify bodies of local community people. He worked long hours for more than a month, with interrupted sleep and constant pressure from community members and CFA colleagues. After a two-week break, he returned to work and, with a physical injury, was placed on light duties. Incredibly, he was directed to work on evidence-gathering from the emergency call tapes for the Victorian Bushfires Royal Commission. He said: "It's one thing to listen to audio tapes. It's one thing to see dead bodies. It's different to actually put the conversation that you're having to the image ... That was the straw that broke the camel's back." In a similar vein, Brad concluded: "I'm sick of it. I've had enough of it. I don't believe the organization anymore ... You virtually end up believing that you're no good at [your job] anyway."

Just as men suffered in the home by not being "the strong one" (Enarson and Fordham, 2001, p. 49), they did not want to appear weak in the workplace. Alternative work roles were rarely an option, and a gradual, supported return to work was not offered. Instead, men told of organizational responses that sought to demote or remove—consequences that are minimized or denied by ESOs listing their employee support services. Such services on offer were often seen as futile and an unlikely setting for disclosure or sharing of feelings with other men.

Health effects

Denial, repression of feelings and constant efforts to live up to Western ideals of manhood can lead to stress, illness and early death (Greig *et al.*, 2000). Along with male privilege come poorer physical and mental health than women, and higher levels of alcohol abuse and loneliness (Kimmel, 2002; Connell, 2005; Kahn, 2010). Drinking alcohol is often part of enacting masculinity in Australian society, and becoming emotional is more accepted in men when they are drunk. After the fires, community meetings were often held in pubs and alcohol was donated. "I wouldn't stop until I was passed out basically, for four weeks ... It was just something that made the pain a little easier," Aaron recalled, following Paul's recollection: "They were using alcohol. They were even using speed ... It was tragic watching them deteriorate." Several men attributed the emergence or worsening of chronic or life-threatening illnesses to Black Saturday, including Parkinson's disease, lymphoma, heart disease, aorta problems, lung and respiratory illnesses. Two developed eating disorders. Four of the 32 men reported they had felt suicidal, with two of these having planned to "take out" others, too. One explained:

> There have been many times where I've certainly thought about ending it all [thinking] "This is just not worth it" or "How many people can I kill?" ... I'm quite happy for it to be on the record, I've felt very suicidal on very many occasions ... To get up in the morning and think to yourself, "Why am I bothering to do this? What's left?" Because there's nothing left.

Most kept these feelings to themselves but another man remembered: "I just felt like hanging myself. So I just sat there ... You are the only two I've told. I've told no-one else."

Reluctance to seek help

Informants richly described the suffering that followed their perceived failure to live up to the prescribed hegemonic male role, not realizing that few men ever do. Only one man didn't cry during his interview and some wept throughout. The men observed other men getting on with things but seeming sad, and not talking about things, not complaining, "suffering in silence," and "keeping it to themselves." Another spoke of "going into his cave." They revealed avoiding speaking of the fires, avoiding conversations altogether and even avoiding community. Many men spoke of managing their inner turmoil by working hard, drinking alcohol or keeping the company of other men in the same circumstances. This collusion from other males and male-dominated emergency response agencies may have contributed, for some men, to the crisis of violence that eventually surfaced. Even four years on, some had become reclusive, some had never regained employment or financial security and many were still steeped in emotions the disaster provoked.

Cook and Mitchell found the "macho" nature of the job and "rescue mindset" of firefighters were obstacles to uptake of services (2013, p. viii). The men recognized they needed help, aware of tension and emotions that couldn't easily be pushed away. One man said, "I could frighten people when they rang, with my emotions ... emotions you can't control." Men told of their belief that they would be judged, stigmatized or thought mentally ill or weak for asking for help. One worker, commenting on an effort to engage fire-affected men, said:

> I saw advertised a month ago there was something about "Are you angry all the time? Come and join a men's self-help group." They would have all just looked at it and gone "Oh fuck off, I'm not interested in that."

Men who did reach out for support often found it lacking. They could wait weeks for an appointment only to find the service they received to be unprofessional and damaging. Luke said, "I think it's all the promising of help and service and then there isn't anything."

Hyper-masculinity and violence

In our earlier research with women after Black Saturday (Parkinson and Zara, 2013), 17 women recounted their partners' and ex-partners' violence following the bushfires, sometimes a new experience and sometimes a sharp escalation building on the past:

> He was a fragile character before, but he was a whole egg. But whatever happened to him through the fires smashed him; whereas a stronger shell might have held, he was smashed, and his moral compass was decimated.
>
> (Cathy)

The suppression of men's feelings—demanded by society in order to support the gender hierarchy—compounds recovery and impairs resilience. Women, too, were silenced, prohibited from speaking about men's violence. Instead, they were encouraged to "give it some time" or reminded that "he's not himself." These were "good" men, "suffering" men, "traumatized" men.

The men described anger as a feature of the aftermath, including in public meetings and in committees established for the reconstruction. "Manly" emotions of acting out, being aggressive or physically provocative, yelling and domineering were observed and acknowledged, as Scott recalled: "I just was ready to drag him across the table at a meeting in public and beat the crap out of him." Indeed, physical violence and aggression is generally associated with masculine behavior, and often rewarded politically and publicly. Perceptions of toughness influence who is considered "leadership material." One relief worker told of a man chasing his children with an axe. Two men spoke of firsthand knowledge of domestic violence they linked to the disaster. Paul referred to his street, "replete with domestic violence" and of his frustration with the lack of intervention from police and community services, noting an eight-week gap between referral and response in one instance. Rod spoke poignantly of his efforts to protect his daughter and grandchildren from his son-in-law:

> The police were called on numerous occasions ... The police were very understanding, much more so than I think he perhaps deserved ... We thought he might ... top the lot of them ... There were times I felt threatened because he's built like a brick toilet and I was always aware that if he did decide to take a swing ... There'd be absolutely no question that he'd flatten me with one punch.

While some men were physically injured, for others there was embarrassment or humiliation in the public realm. "Mob mentality," "harassment" and "bullying" were used to describe anger in community meetings by "big burly men, shouting at people." Steve explained: "I saw people behaving badly, really being abusive, domineering ... Men are combative, competitive, aggressive, by nature ... There's a lot of horizontal violence since the bushfire."

Adverse and violent reactions may have come from men for whom "a high degree of personal and interpersonal control is important" (Houghton, 2009, p. 101). It may begin with a stubborn refusal to leave the home as fires near, even though danger is revealed to be inevitable and evacuation is logical (Tyler and Fairbrother, 2013; Parkinson, 2015). A man's status as head of the household and decision-maker led to the deaths of individual men whose families had left, or couples and whole families, as Brad stated:

> I have first-hand knowledge that there are women, wives, on Black Saturday who wanted to leave town and their husband said, "No, we're staying to fight this." And they stayed to fight and they both died.

Gendered vulnerabilities and risks: recommendations for change

The narratives of the 32 men in this sample attest to the damage of strict gender roles. It was manifestly impossible for men to protect and provide in the face of the catastrophic Black Saturday bushfires. Yet there was, and remains, no interrogation of the notion of masculinity or our expectations of men. Four years on, individual men struggled to understand the assault. Seven years on, the struggle for many continues, but social expectations of gendered roles have not shifted. It appears that as each year passes, less is remembered about the inadequacy of the way we construct gender roles, falsely assuming men protect and women are protected. It appears that silencing suffering men, demoting them or removing them are easier than addressing the fallacy of the gender dichotomy. It is time to take seriously Luke's question:

> One of my workers … was talking about his life and his parents and blah, blah, blah, and how bad he felt. Life wasn't worth living and all this. And I thought, "I don't need this." That was my first reaction because I'm a bloke … Why don't we talk about these things? Why do you have to wait until you're bloody 50, pulling your hair out and then it all turns to shit and there's yelling and screaming?

The privilege attached to being male comes at a cost—and disasters show this to be very high. Just as socially constructed patterns of masculinity have historically positioned men in the frontline for harm during disaster in Australia, the designated role for women brings different risks. It is critical to note that females accounted for 42% of those who died on Black Saturday, many of these while evacuating alone or with children. Gendered analysis shows that a similar proportion of bushfire deaths was recorded over the previous 50 years: 40% female and 60% male (Haynes *et al.*, 2008). Christine Eriksen writes that gendered expectations of men and women alike shape their vulnerability in disasters and that "men … often take control and perform protective roles that many have neither the knowledge nor the ability to safely attempt to fulfill" (2014, p. 39).

Broadening the range of acceptable behaviors for men and women is essential. It is critical, however, that men take on their share of domestic and caring labor before women are tasked with further work in disasters (McLennan and Birch, 2006; Branch-Smith and Pooley, 2010). The risk is that women would be burdened with further expectations while men are relieved of theirs. A related risk is that men will continue to hold the majority of leadership positions in Australian emergency management and government, and among voluntary firefighters (McLennan and Birch, 2006).

This research (and see Parkinson and Zara, 2013; Parkinson, 2015) suggests that positive change involves identifying points of vulnerability and reshaping our understanding of how men and women "should" behave in and after a disaster. The aim of these recommendations is to reduce the compounding effects of gender on disaster impacts and expand the range of acceptable behaviors for both women and men. This begins with four strong steps: reduce gender stereotyping; reduce vulnerability of emergency services workers and other first responders; improve individual support for survivor physical, mental and emotional health; and offer equal opportunities and respect to all disaster survivors.

In Australia, one of the legacies of the horrific Black Saturday fires is a path-breaking new Gender and Disaster (GAD) Taskforce, created by Emergency Management Victoria in collaboration with Women's Health Goulburn North East and Monash University Disaster Resilience Initiative. The taskforce engages high level representatives from all major Victorian emergency service organizations, the community, government, nongovernment, academic and women's health sectors. The purpose of the GAD Taskforce is to provide "statewide strategic direction and leadership to reduce the compounding effects of gender on disaster impacts" (Victorian Emergency Management, p. 1).

The work of the Taskforce aims to bring a gender focus to disaster policy, planning, training and practice in order to improve the support that men and women receive before and after disaster, mitigate risks to men's and women's health and well-being post-disaster, and build awareness of the critical need for attention to gender in disaster planning and community recovery. As gender inequality and societal complicity contribute to the prevalence of men's violence against women, changing the culture of emergency management to include equal contribution by women at all levels has been identified as an important part of this strategy.

Devoted to moving these aims from the abstract to the specific, the first steps are underway to recognize and reduce gender harms to men and boys, women and girls in the floods, fires and other disasters certain to affect the nation in future years.

References

Atkins, C. (2011) Aftershock: The ongoing impact of disasters. *Emergency Management Insight*, 5, pp. 4–7.

Branch-Smith, C. and Pooley, J.A. (2010) Women firefighters' experiences in the Western Australian Volunteer Bush Fire Service. *Australian Journal of Emergency Management*, 25(3), pp. 12–18.

Cameron, P., Mitra, B., Fitzgerald, M., Scheinkestel, C., Stripp, A., Batey, C., Niggemeyer, L., Truesdale, M., Holman, P., Mehra, R., Wasiak, J. and Cleland, H. (2009) Black Saturday: The immediate impact of the February 2009 bushfires in Victoria, Australia. *Medical Journal of Australia*, 191(1), pp. 11–16.

Cartwright, T. (2012) Conference opening. In: Proceedings of the Hidden Disaster Conference, Melbourne, Australia: Women's Health in the North, Women's Health Goulburn North East and Australian Domestic and Family Violence Clearinghouse. Available from http://www.whin.org.au/what-we-do/hidden-disaster-conference.html (accessed December 10, 2015).

Connell, R.W. (2005) *Masculinities*. 2nd edn. Los Angeles, California: University of California Press.

Cook, B. and Mitchell, W. (2013) *Occupational health effects for firefighters: The extent and implications of physical and psychological injuries*. Victoria, Australia: United Firefighters Union of Australia.

Cox, H. (1998) Women in bushfire territory. In: Enarson, E. and Morrow, B.H. (eds.) *The gendered terrain of disaster: Through women's eyes*. Westport, CA: Praeger Publishers, pp. 133–142.

Desmond, M. (2008) The lie of heroism. *Contexts*, 7(1), pp. 56–58.

Elinder, M. and Erixson, O. (2012) Gender, social norms, and survival in maritime disasters. *Proceedings of the National Academy of Sciences*, 109(33), 13220–13224. Available from doi: 10.1073/pnas.1207156109 (accessed April 15, 2015).

Enarson, E. (2006) Women and girls last: Averting the second post-Katrina disaster. Available from http://understandingkatrina.ssrc.org/Enarson/ (accessed December 10, 2015).

Enarson, E. and Fordham, M. (2001) Lines that divide, ties that bind: Race, class, and gender in women's flood recovery in the US and UK. *The Australian Journal of Emergency Management*, 15(4), pp. 43–52.

Eriksen, C. (2014) *Gender and wildfire: Landscapes of uncertainty*. New York: Routledge.

Glaser, B. and Strauss, A. (1967) *The discovery of grounded theory: Strategies for qualitative research*. Chicago: Aldine.

Greig, A., Kimmel, M. and Lang, J. (2000). *Men, masculinities and development: Broadening our work towards gender equality*. Gender in Development Monograph Series #10. Available from http://www.engagingmen.net/files/resources/2010/Caroline/Men_Masculinities_and_Development.pdf (accessed December 10, 2015).

Haynes, K., Tibbits, A., Coates, L., Ganewatta, G., Handmer, J. and McAnerney, J. (2008) *100 years of Australian civilian bushfire fatalities: Exploring the trends in relation to the "stay or go policy."* Commissioned by the Bushfire Co-operative Research Centre. Sydney, Australia: Risk Frontiers.

Hoffman, S. (1998) Eve and Adam among the embers: Gender patterns after the Oakland Berkeley firestorm. In: Enarson. E. and Morrow. B.H. (eds.) *The gendered terrain of disaster: Through women's eyes*. London: Praeger, pp. 55–61.

Houghton, R. (2009) "Everything became a struggle, absolute struggle": Post-flood increases in domestic violence in New Zealand. In: Enarson, E. and Chakrabarti, P.G.D. (eds.) *Women, gender and disaster: Global issues and initiatives*. New Delhi: Sage, pp. 99–111.

Johnson, J.L. and Repta, R. (2012) Sex and gender: Beyond the binaries. In: Oliffe. J.L. and Greaves. L. (eds.) *Designing and conducting gender, sex, and health research*. Thousand Oaks, CA: Sage Publications, pp. 17–39.

Kahn, J.S. (2010) Feminist therapy for men: Challenging assumptions and moving forward. *Women and Therapy*, 34(1–2), pp. 59–76.

Kimmel, M. (2002) Toward a pedagogy of the oppressor. *Tikkun Magazine*, 17(6), pp. 42–45.

Manne, R. (2009, July) Why we weren't warned? The Victorian bushfires and the Royal Commission. *The Monthly*. Available from https://www.themonthly.com.au/

monthly-essays-robert-manne-why-we-weren-t-warned-victorian-bushfires-and-royal-commission-1780 (accessed December 10, 2015).

McFarlane, A. (2005) Psychiatric morbidity following disasters: Epidemiology, risk and protective factors. In: López-Ibor, J.J., Christodolou, G.G., Maj, M.M., Sartorious, N.N. and Okasha, A. (eds.) *Disasters and mental health*. Chichester, West Sussex: John Wiley and Sons, pp. 37–63.

McLennan, J. and Birch, A. (2006). Survey of South Australian Country Fire Service Women Volunteers: South Australian Country Fire Service Report Number 2006: 1. Bushfire CRC Enhancing Volunteer Recruitment and Retention Project (D3) [Book Review]. *Australian Journal on Volunteering*, 11(2), p. 81.

Pacholok, S. (2009). Gendered strategies of self: Navigating hierarchy and contesting masculinities. *Gender, Work and Organization*, 16(4), pp. 71–500.

Pacholok, S. (2013). *Into the fire: Disaster and the remaking of gender*. Toronto: University of Toronto Press.

Parkinson, D. (2015) Women's experience of violence in the aftermath of the Black Saturday bushfires. PhD. Monash University.

Parkinson, D., Lancaster, C. and Stewart, A. (2011) A numbers game: Women and disaster. *Health Promotion Journal of Australia*, 22(3), pp. 42–45.

Parkinson, D. and Zara, C. (2013) The hidden disaster: Violence in the aftermath of natural disaster. *The Australian Journal of Emergency Management*, 28(2), pp. 28–35.

Pease, B. (2010) *Undoing privilege: Unearned advantage in a divided world*. London: Zed Books.

Pease, B. (2014) Hegemonic masculinity and the gendering of men in disaster management: Implications for social work education. *Advances in Social Work and Welfare Education*, 16(2), pp. 60–72.

Proudley, M. (2008) Fire, families and decisions. *The Australian Journal of Emergency Management*, 23(1), pp. 37–43.

Rivers, J. (1982) Women and children last: An essay on sex discrimination in disasters. *Disasters*, 6(4), pp. 256–267.

Scanlon, J. (1997) Human behaviour in disaster: The relevance of gender. *Australian Journal of Emergency Management*, 11(4), pp. 2–7.

Spradley, J.P. (1980) *Participant observation*. Fort Worth: Harcourt Brace Jovanovich College Publishers.

Tyler, M. and Fairbrother, P. (2013) Gender, masculinity and bushfire: Australia in an international context. *The Australian Journal of Emergency Management*, 28(2), pp. 20–25.

Victorian Bushfires Royal Commission (2010). *Final Report: Summary*. Melbourne: Victorian Government. Available from http://www.royalcommission.vic.gov.au/final documents/summary/PF/VBRC_Summary_PF.pdf (accessed December 22, 2015).

Victorian Emergency Management (2014) Gender & Disaster Taskforce, Terms of Reference, available from http://www.whealth.com.au/documents/environmentaljustice/ GADT_TermsOfReference.pdf (accessed March 22, 2014).

Zara, C. and Parkinson, D. (2013) *Men on Black Saturday: Risks and Opportunities for Change*. Wangaratta: Women's Health Goulburn North East. Available from http://www.whealth.com.au/documents/work/about-men/Men-on-BS-Report.pdf (accessed December 22, 2015).

8 The tsunami's wake

Mourning and masculinity in Eastern Sri Lanka

Malathi de Alwis

On December 26, 2004, a tsunami of unprecedented proportions decimated two-thirds of Sri Lanka's coastline, densely populated with primarily lower-income housing as well as many hotels and guesthouses located along the southern and eastern coasts. It left 33,000 dead, 4,000 missing and 21,000 injured. It reduced 98,000 houses to rubble and damaged even more, leaving over a million people displaced.

The Muslim-dominated, densely populated coastal areas of the Eastern Province, the first point of landfall for the tsunami, sustained the greatest amount of devastation with 10,000 dead and more than ten times that number rendered homeless. Both the Eastern and Northern Provinces peopled primarily by minority populations of Tamils and Muslims experienced a disproportionate percentage of death and displacement while reeling from over two decades of civil war that had already decimated families and destroyed their homes and lands.

The tsunami killed three times more women than men and more than a third of the victims were children, thus irrevocably transforming family structures and kin networks; in some instances, entire families were wiped out and it was not unusual to hear people saying that they had lost sometimes 30, 56 or 82 relatives within 15 minutes. The death of an unprecedented number of women introduced a new demographic in the war zones of the north and east with a large number of widowers and young, single fathers joining the extensive ranks of widows and young, single mothers in these regions. Many of the widowers married within a year after the tsunami, often citing difficulty in looking after their children by themselves. Widows, on the other hand, have been reluctant to enter into a second marriage, fearful that their children would be ill-treated by their stepfathers (Hyndman, 2009).

The acute grief, constant mental trauma and daily suffering the tsunami engendered defy even these staggering series of enumerations. We can only briefly sense it, perhaps, when we sit with 60-year-old Yogendran in his brand-new, empty, two-story house in Navalady, Batticaloa, in the Eastern Province, listening to melancholic Tamil film songs which he plays every day at dusk, as light bounces hollowly off the photographs of his wife, four sons, mother-in-law, father-in-law, sister-in-law and best friend, in the shrine he has built to his deceased family. Sixty-five-year-old Sellamuttu's loss is painfully palpable as he lovingly brushes the dust off the imprint left by his daughter, then a toddler, when she accidentally stepped onto the kitchen floor that was being cemented. That tenderly inscribed foundation is all that remains of his old house, and his daughter, who died along with her husband and two children as well as her mother. Sellamuttu now lives alone in his brand-new, empty, two-story house that abuts the foundation of his former house, about half a kilometer down the road from Yogendran. None of Sellamuttu's three sons, who all fortunately survived the tsunami, wishes to return to his village, especially now that they have young families of their own. In fact, there is barely a child to be seen in these picturesque villages, sandwiched between the ocean and the lagoon, now peopled predominantly by aging men. Yet Sellamuttu insists that they are very much present, in spirit, and that their joyful laughter keeps him company at night. He cites this as one of the primary reasons he will not abandon his village and move in with his sons, who live in town.

One could shrug off such statements, as many are wont to do, as the sorrowful fancies or drunken hallucinations, or both, of a man who is unable to "move on" with his life or "come to terms" with his grief. Indeed, alcoholism is not uncommon among these men, several of them having frittered away all their compensation money on cheap, illicit alcohol. Many NGO practitioners, on hearing me speak of men such as Yogendran and Sellamuttu, have been quick to comment that if they had had proper "psychosocial support" they would no longer need to cling to spirits—of either kind.

This chapter seeks to disrupt and decenter a particular formulation of Tamil masculinity—alcoholics and wife beaters—that has been produced through humanitarian and development aid discourses in Sri Lanka in the wake of the December 2004 tsunami, a discourse I have critiqued elsewhere, on other grounds (de Alwis 2009b, 2015). It has been well documented that alcoholism and domestic violence did increase in the aftermath of the tsunami, especially in the camps for those displaced by it (APWLD, 2005; de Alwis, 2005; Fisher, 2005; van der Veen and Somasunderam, 2006). What I wish to disrupt here is the assumption that this was a predominant form of masculinity, along with that of the ruthless yet heroic militant, in Sri Lankan Tamil society. I wish to argue that there were other, less negative masculinities and masculinized practices that also came to the forefront during this period but were rendered invisible by hegemonic narratives. Many men wrestled daily with overwhelming grief and loss, and, in a context where the gender demographics were up-ended by so many more women than men dying, they struggled in their new role of primary caregiver. Many men also initiated and participated in designing and building both private and collective memorials to their deceased kin. By tracing the fragility and precariousness of the latter enterprise, I hope to push the parameters of humanitarian/development aid discourses to take cognizance of the fact that masculinity is an "ambivalent complex of weakness and strength" (Chopra *et al.*, 2004, p. 8) which involves "an interplay of emotional and intellectual factors" that directly implicates women as well as men and is mediated by other social factors such as ethnicity, religion and class (Berger *et al.*, 1995, p. 3).

"Dangerous individuals"

Neither Yogendran nor Sellamuttu received psychosocial support despite the existence of many international, national and local NGOs that offer such assistance in the Batticaloa district. This reflects a central assumption underlying such therapeutic interventions, though frequently not articulated as such, that women and children are most in need of psychosocial therapy. Vanessa Pupavac, who makes a similar observation, notes that such a "bias" is prevalent "irrespective of whether men in the target recipient community have actually been exposed to greater distress as soldiers on the front line or prisoners of war" (2001, p. 363). This bias also ignores the fact that women and children are more capable of adapting to their circumstances: "Women may maintain purposefulness in their traditional role as primary family carers. Children can escape into play in even the most adverse situations or, in more fortunate circumstances, may become integrated into new communities through schooling" (p. 364).

I suggest such an assumption—women's and children's greater need for psychosocial therapy—stems from a prior sociological assumption that has been made about societies in which these interventions are most often conducted; that is, being a woman or a child is in and of itself a psychological hardship due to the marginalized and vulnerable status inhabited by both. The flip side of such an assumption of course is that it is men who are always dominant and aggressive. Among NGO practitioners, this "always already" notion of masculinity was further concretized through post-tsunami increases in alcoholism and domestic violence, as noted above. This conceptualization of masculinity is also buttressed by a psychosocial model of war-torn/disaster-prone societies as being susceptible to traumatic symptoms that cause dysfunctionalism, which in turn may lead to abuse and violence (Pupavac, 2001). International mental health consultants for WHO, UNICEF and other relief agencies, notes Summerfield, "claim that early intervention can prevent mental disorders, alcoholism, criminal and domestic violence, and new wars in subsequent generations by nipping 'brutalization' in the bud" (1997, p. 1568). Alcoholics Anonymous groups and drug rehabilitation programs clearly sit well with this kind of rationalization as they target "dangerous individuals" who usually tend to be men. Most often, such interventions are framed within the contours of "addiction" and "violence" rather than "suffering" or "mourning" or "melancholia," and are primarily focused on "treating" a very specific "character defect" or aspect of a man's persona.

The contours of mourning and melancholia

In Sigmund Freud's now classic essay of 1917, "Mourning and Melancholia," he argues that the process of mourning enables a working through of grief by gradually severing the libido's attachment to the lost object. He juxtaposes this state of being with melancholia, an aberrant form of mourning, where the ego denies the loss of the loved object and withdraws into itself while simultaneously withdrawing a memory trace of the lost object into itself (1948). Such a process is charged with ambivalence as the ego both rages and loves, hates and desires itself as well as the love object. Sometimes this state of torment can transpose into mania which is accompanied by a completely different symptomology of high spirits and uninhibited action, though the melancholic does not realize that any change has taken place in him (Freud, 1948, p. 155). Ironically, Freud likens alcoholic intoxication to this same group of conditions: "a relaxation produced by toxins of the expenditure of energy in repression" (p. 165).

The distinguishing mental features of melancholia are "a profoundly painful dejection, abrogation of interest in the outside world, loss of the capacity to love, inhibition of all activity, and a lowering of the self-regarding feelings to a degree that finds utterance in delusional expectation of punishment" (Freud, 1948, p. 153). Interestingly, the same traits are met with in grief and mourning, with one exception: the fall in self-esteem is absent in grief. "It is easy to see," notes Freud,

> that this inhibition and circumscription in the ego is the expression of an exclusive devotion to its mourning, which leaves nothing over for other purposes or other interests. *It is really only because we know so well how to explain it that this attitude does not seem to us pathological.*
>
> (p. 153, emphasis added)

I am emphasizing this final sentence as especially prescient given the heated debates that currently coalesce around the classifying of those who have experienced unimaginable horrors, as "traumatized," rather than as suffering or distressed or grieving (for a discussion of some of these debates, see de Alwis, forthcoming).

Melancholia is also an unconscious process for "he knows whom he has lost but not *what* it is he has lost in them," unlike that of mourning "where there is nothing unconscious about the loss" (Freud, 1948, p. 155, emphasis in original). However, Judith Butler, in a thoughtful rereading of Freud's essay, has argued that the temporal logic of Freud's own terms leads to an undoing of the distinction between "normal mourning" and melancholia because the inward-turning ego could not be said to securely preexist the lost object. "Melancholia defined as the ambivalent reaction to loss, may be coextensive with loss so that mourning is subsumed in melancholia" (Butler, 1997, p. 174, see also Jeganathan 2008). Such a re-reading illuminates the "work of melancholia" that produces the psyche as a distinct domain but cannot obliterate the social occasion of its production. It is through this ambivalent interplay between the social and the psychic, mourning and melancholia, life and death that I seek to offer my own, brief reading of three different unfoldings of masculinized mourning that struggle with extraordinary loss.

"Sipping his way to oblivion"

Daya Somasunderam's work on the psychological impact of the civil war on Tamil civilians in the Northern Province offers a haunting case study of what he terms a "Grief Reaction." A middle-aged engineer, whose three children and mother-in-law had been pulled out of their home and shot dead on the street "for no apparent reason" by the Indian Peace Keeping Forces (IPKF), suffered recurrent nightmares about the suffering they had undergone. He was particularly anguished about his pretty little daughter, whose frock had been lifted up by a soldier and shot through the groin. She had had to drag herself on the road and had bled to death due to the lack of medical attention (Somasunderam, 1998, p. 233). This engineer spent his days "in deep sorrow" alternating between seeking to avenge the deaths of his children—"I will personally kill these soldiers"— and bouts of crying (p. 233). "Previously an occasional drinker, he had now started drinking heavily and was in a state of intoxication most of the day and night" (p. 234).

I often recalled this nameless engineer when I encountered Sivanesan, a middle-aged fisherman who was Sellamuttu's neighbor. The first time I met Sivanesan was when he stopped by Sellamuttu's house, his sarong hiked up over his knees, swaying from side to side like a coconut tree in a storm and announcing to all and sundry that he was trying to find some insecticides to put an end to himself. During more sober, lucid moments, Sivanesan and I would discuss the texture of the goat hide he was stretching across a drum (a very caste-marked occupation), the lack of women and children in the neighborhood and his desire for, yet inability to find, a second wife.

It was Sellamuttu who informed me that Sivanesan had lost his wife and two daughters in the tsunami. Only his eldest daughter had survived, though she had lost her husband and two babies. Sivanesan had managed to arrange a second marriage for her and she had got pregnant right away unlike many other women, such as one of Sellamuttu's daughters-in-law, who had ingested "the black water" and become infertile. Though the baby boy was somewhat sickly, this new life amidst all the devastation was considered to be a good omen until disaster struck again. Sivanesan's daughter died due to complications that arose during the birth of her second son. That was the day Sivanesan had come in search of insecticides.

This fresh tragedy precipitated discussions of what people perceived to be Sivanesan's greatest and original tragedy. Four years before the tsunami, Sivanesan's only son had been forcibly conscripted by the Liberation Tigers of Tamil Eelam (LTTE) while he was cycling home from school. Sivanesan had spent many months tirelessly searching for him, pursuing whatever tenuous lead he could ferret out, to no avail. Three years later, the LTTE had suddenly handed Sivanesan a certificate attesting to the fact that his son had died in battle.

After his daughter's death, Sivanesan's drinking got steadily worse and he started "becoming a nuisance" in the village, singing loudly at night and picking quarrels with neighbors. Things finally came to a head one night when Sivanesan tried to stroke the head of a young woman on her way back home, apparently mistaking her for his daughter, and he was chased out of the village. A prediction made by one villager, some years previously, that Sivanesan was "sipping his way to oblivion" seemed to have been proved correct.

"The amman is now my protector"

Imbued with ecstatic devotion, Murugaiah was shaking and spinning and lunging at the crowds of devotees who thronged the Kannaki Amman temple, when two women standing near me, who turned out to be his wife and daughter, began speaking about his miraculous transformation.

Murugaiah's son had been shot in front of him, along with his cousin, during a Sri Lankan army cordon and search operation. This boy had been the apple of his eye. Murugaiah had devised an ingenious way to hide him from the LTTE conscriptors because he could not bear to send him away to Colombo. As a result, Murugaiah blamed himself for his son's death. After the funeral, he had stopped going out to work as a day laborer and gradually lapsed into silence, barely noticing the rest of his family or what was going on around him and "eating as if he could no longer taste his food." His daughter's eyes filled with tears when recounting this "time of sadness" when her father was lost to her.

Murugaiah's condition was immediately recognizable to those around him as the affliction of *tanimai*, glossed as "aloneness" by anthropologist Valentine Daniel in his Piercian analysis of this state of malaise (1989, p. 69). Its symptoms are very similar to those of melancholia discussed above as it is experienced unself-consciously and involves being "disconnected from other human beings with whom one ought to be connected" (p. 78). While thus afflicted, Murugaiah suffered a bad attack of chicken pox, popularly understood to be an illness brought on by the amman (mother goddess). Murugaiah's wife and daughter were visiting him in hospital when the tsunami struck. His mother and younger daughter were alone at home and were swept away along with their house. Their bodies are missing to this day.

Murugaiah's wife and daughter decided not to tell him what had happened to their house or the rest of his family, until he was discharged from hospital. But, when they next went to visit him, Murugaiah was fully aware of their calamity. He professed that Kannaki Amman had appeared in a dream to him and said: "You have lost all but I will always protect you. Go and look after the two you have remaining." Murugaiah returned to his village a new man. He rebuilt their home single-handedly and now even works on weekends in order to collect money for his daughter's dowry. Whatever free time he has he spends at the Kannaki Amman temple in his village and frequently goes into a trance (a sign that the amman has entered his body) during special festivals at the temple. Murugaiah still doesn't speak much; though he did confide in me once: "I couldn't touch my mother's feet [a sign of respect] before she died... The Amman is now my protector."

"Generations to come must remember"

Sellamuttu was one of the first people to return to his village after the tsunami. As noted at the beginning of this chapter, he lost his wife, daughter, son-in-law and several grandchildren. Though his three sons, who survived, constantly urge him to come and live with them, he keeps rejecting their invitations saying he is happiest where he is closest to the dead. He has tried nine times to keep a dog to assuage his loneliness but each puppy has wasted away and died after a few months, a sign that there are too many troubled spirits in the neighborhood, according to Sellamuttu. He built a small extension to his house and rented it out to a long-distance bus driver but when that man returns to his quarters he is too tired to chat with him and goes straight to bed. Sellamuttu's chief companions are a few other aged men in the village and his three sons, who stop by to check on him regularly. Most of his other relatives are too scared to visit, he noted forlornly, as they are fearful that another tsunami could envelop his new house. Nonetheless, he commands great respect in his village and many seek his advice and guidance.

After supervising the building of his new house (funded by a Christian INGO), Sellamuttu focused his energies on building a large, public tsunami memorial facing the sea, beside the coastal road abutting his village. Because the proposed site was on state land, Sellamuttu had had to expend a great deal of energy to get permission to build on it. Work was begun a year after the tsunami as he and the rest of his village felt it was imperative that "generations to come must remember this great tragedy that befell our village." Sellamuttu is very proud of the fact that the entire enterprise was supported by financial contributions as well as labor from his village, and that "every family in the village made a donation, even if all they could afford was Rs 100." The Memorial Building Committee also jointly composed the poem that appears on the granite panels along with the seemingly endless list of the names of the dead.

Sellamuttu faithfully switches on the lights that illuminate the memorial every night, when he returns home with his meager meal comprised of offerings to a Hindu deity at a temple in a neighboring village. However, this nightly ritual will probably cease soon as it is only a matter of time before the Electricity Board shuts off the electric supply due to nonpayment of bills. The Memorial Building Committee's coffers are empty and there is talk of misappropriation of funds, but Sellamuttu's greatest worry is that there will be no one to continue maintaining the monument when those of his generation die out. The last time I met him, he talked enthusiastically about his new project to build a fence around the memorial and urged me to help him find funds for it.

Sivanesan's, Murugaiah's and Sellamuttu's lives reflect three very different trajectories. All three have struggled to live with extraordinary loss and grief: Sivanesan has sought solace in alcohol, Murugaiah in religion and Sellamuttu in memorial building. While Sivanesan's alcoholism eventually led to his ostracization and eviction from his village, Murugaiah and Sellamuttu have succeeded in reconnecting with society, in different and distinctive ways. While the resort to religious devotion and spirit possession to deal with extraordinary loss and grief has been substantively explored by several anthropologists working on Sri Lanka (de Alwis 1997, Lawrence, 1999; de Alwis, 2001; Perera, 2001), the resort to memorialization has received less attention (Jeganathan, 2008, 2010; Simpson and de Alwis, 2008; de Alwis 2009a, 2009b). In the section that follows, I wish to ponder more deeply about the building of memorials as a work of communal mourning and memorialization.

Communal mourning and memorialization

Sellamuttu's public endeavor of memorial building is clearly apprehensible under the category of "community activism" or "community welfare work." The notion of "community" at play here is a complicated one; it is an articulation of colonial and postcolonial reformulations that Pradeep Jeganathan (2009) has usefully traced for us. British colonial administrators, Jeganathan argued, sought to transpose English village structures onto Sinhala villages, guiding them into a "patriarchal system" led by the "natural" leaders of the "community" (p. 67). Nationalist revivalists and reformists such as Ananda Coomaraswamy and Anagarika Dharmapala also contributed to subsequent romanticizations of "village community," leading to the formation of Village Protection Societies and Rural Development Societies across the island, thus actively engaging with, appropriating and indigenizing colonial ideas and policies (pp. 70–71).

Sellamuttu as well as Vishnupillai and Balendran, livewires in the neighboring village tsunami memorial committees, used to be members of their respective Rural Development Societies and Kovil [Hindu temple] committees but they have "no taste" for that any more. In fact, Vishnupillai swears he will not "lift a finger" to help rebuild the Murugan Kovil as he has lost all faith in religious deities, after his entire family was wiped out by the tsunami.

"Community" is a ubiquitous term today. No tsunami reconstruction project proceeds without "community participation" and "community consultation." However, as Jeganathan (2007) has observed, there is a universalization of "voice" in such processes while de Alwis (2009b) has criticized the assumption that such a "voice" is transparently available through focus group discussions or Participatory Rural Appraisal (PRA). Community, it must also be remembered, is hierarchically and politically constituted and always exclusionary (de Alwis 2009b). Indeed, the "community" which was involved in building memorials, along the eastern coast, from Batticaloa to Thirukkovil, was comprised solely of men. Most groups comprised Hindu Tamil men, such as in Sellamuttu's, Vishnupillai's and Balendran's groups, the exception being the Dutch Bar memorial committee, which was composed of Tamil-speaking Burgher men (descendants of Portuguese and Dutch colonizers) who were Catholics.

The collective memorials are now an organic part of the post-tsunami landscape of Eastern Sri Lanka. They are not merely markers of the sensitivity and largesse of each village collective whose members generously contributed to erect them irrespective of whether they had lost someone in the tsunami or not. They are also not merely the markers of the sweat and toil of the memorial committee members who not only planned and coordinated the work but also often labored to build the memorials. Neither are they merely markers of those who lost their lives so tragically in the tsunami. They are also reminders to all those who participated in designing and building them and others who merely observed the process from afar of much strife and negotiation and compromise as a variety of male egos battled it out on the turf of tsunami memorialization.

There were several occasions when I would seek out the leader of a memorial committee to discover that he had been summarily thrown out and replaced by a rival. Committee members often accused each other of rampant corruption and of bowing to political pressure, the latter aspect coming to the forefront during annual tsunami commemorations when heated debates would take place regarding which local dignitaries and militant groups should be invited to the event. Noncompliance with militant diktats could turn violent as one memorial committee discovered when all the bulbs on their memorial were smashed one night "by unknown elements." However, this long-drawn-out process of domination and subordination, fisticuffs and embraces, failures and triumphs, I would argue, did enable a certain working through of loss, grief and suffering for a majority of the men involved in this process.

Conclusion

I have offered a multilayered glimpse into a particular post-tsunami context on the east coast of Sri Lanka in order to illuminate how a reading of that context must contend with a variety of assumptions about gender, humanitarian aid, trauma, power and grief—and turn them on their heads. The tsunami not only devastated lives and livelihoods but also social relationships and ways of being. It also engendered certain humanitarian aid discourses that often reinforced gendered stereotypes about "traumatized" Third World peoples. I have sought here to problematize resorting to such stereotypes and to probe more deeply into different kinds of masculinities that struggled with sorrow and loss in very different ways. In this regard, I found the building of public tsunami memorials of particular interest as they seemed to enable men's "work of melancholia" in productive ways.

Memorialization is a familiar process in Sri Lanka. Its landscape is strewn with myriad statues, buildings, parks and streets commemorating British colonial administrators, nationalist patriots, local politicians, generous philanthropists and fallen soldiers. However, what made these tsunami memorialization processes so striking to me, in addition to the monumentality or uniqueness of the design of some, was that their very abundance marked the absence of any civilian-initiated, collective memorials to the greater number of civilians who died in the civil war. The primary reason for this, I would argue, is that "natural" disasters have a very different moral framing than man-made disasters such as civil wars. Disasters are often exacerbated by human interventions such as deforestation, sand and coral mining, and settlement in coastal conservation areas, as was the case in tsunami-affected Sri Lanka. However, the majority of the Sri Lankan populace perceives disasters as "acts of God" that lay waste to lands and peoples irrespective of ethnicity, class, caste, religion or gender, while those who die in civil wars are rarely perceived as innocent. Thus, to commemorate their death(s) publicly was surely to bring on the wrath of either the Sri Lankan state or Tamil militant groups.

I recall vividly the terror etched on the faces of some of my friends who were members of a small women's group in Batticaloa, as they described how they would creep out at the dead of night to paint roads or place markers in public spaces where civilians had been killed, either by government forces or militants. The only permanent public memorials that remain in the Batticaloa region today are a few to slain comrades built by the breakaway faction of the LTTE who aligned with the Sri Lankan government and two that were built by the LTTE to commemorate Sri Lankan military massacres of civilians in Sathurukondan and Kokkadicholai. Surprisingly, these two memorials remain despite the systematic bulldozing of all LTTE-built monuments, memorials and cemeteries in the Northern Province (see de Alwis 2010, 2015).

References

Asia Pacific Forum on Women Law and Development (APWLD) (2005) Why are women more vulnerable during disasters? Violations of women's human rights in the tsunami aftermath. Bangkok: APWLD. Available from http://iknowpolitics.org/sites/default/files/tsunami_report_oct2005.pdf (accessed November 10, 2015).

Berger, M., Wallis, B. and Watson, S. (eds.) (1995) *Constructing masculinity*. London: Routledge.

Butler, J. (1997) *The psychic life of power*. Stanford: Stanford University Press.

Chopra, R., Osella, C. and Osella, F. (eds.) (2004) *South Asian masculinities*. Delhi: Kali for Women.

Daniel, V. (1989) The semeiosis of suicide in Sri Lanka. In: Lee, B. and Urban, G. (eds.) *Semiotics, self and society*. New York: Mouton de Gruyer, pp. 69–100.

de Alwis, M. (1997) Motherhood as a space of protest. In: Basu, A. and Jeffrey, P. (eds.) *Appropriating gender: Women's activism and the politicization of religion in South Asia*. London: Routledge, pp. 185–201.

de Alwis, M. (2001). Ambivalent maternalisms: Cursing as public protest in Sri Lanka. In: Meintjes, S., Pillay, A. and Turshen, M. (eds.) *The aftermath: Women in post-conflict transformation*. London: Zed Books, pp. 210–224.

de Alwis, M. (2005) Responding to sexual and gender-based violence post-tsunami. (Report). Colombo: UNICEF.

de Alwis, M. (2009a) "Disappearance" and "displacement" in Sri Lanka. *Journal of Refugee Studies*, 22(3), pp. 378–391.

de Alwis, M. (2009b) A double wounding? Aid and activism in post-tsunami Sri Lanka. In: de Alwis, M. and Hedman, E. (eds.) *Tsunami in a time of war*. Colombo: International Centre for Ethnic Studies, pp. 121–138.

de Alwis, M. (2010) Sri Lanka must respect memory of war. Available from http://www.guardian.co.uk/commentisfree/2010/may/04/sri-lanka-must-respect-war-memory (accessed November 3, 2015).

de Alwis, M. (forthcoming) Trauma, memory, forgetting. In: Amarasingam, A. and Bass, D. (eds.) *Sri Lanka: The struggle for peace in the aftermath of war*. London: Hurst, in press.

Fisher, S. (2005) Gender based violence in Sri Lanka in the aftermath of the 2004 tsunami crisis. MA. University of Leeds.

Freud, S. (1948 [1917]) Mourning and melancholia. In: Ernest Jones (ed.) *Collected papers, Volume IV*. London: Hogarth Press, pp. 152–170.

Hyndman, J. (2009) Troubling "widows" in post-tsunami Sri Lanka. In: de Mel, N., Ruwanpura, K.N. and Samarasinghe, G. (eds.) *After the Waves*. Colombo: SSA, pp. 17–41.

Jeganathan, P. (2007) Philanthropy, political economy, psyche: Post tsunami reconstruction in a Sri Lankan community. Paper presented at Research Methodologies Conference, Centre for International Studies, University of Science, Penang.

Jeganathan, P. (2008) Ananda, Hamlet, Abdul Majeed: Mourning, melancholia and men. Paper presented at Interrogating Masculinity in South Asia Conference, Rockefeller Centre, Bellagio.

Jeganathan, P. (2009) "Communities": East and West. In: de Alwis, M. and Hedman, E. (eds.) *Tsunami in a time of war*. Colombo: International Centre for Ethnic Studies, pp. 59–81.

Jeganathan, P. (2010) In the ruins of truth: The work of melancholia and acts of memory. *Inter-Asia Cultural Studies Journal*, 11(1), pp. 6–20.

Lawrence, P. (1999) The changing amman: Note on the injury of war in Eastern Sri Lanka. In: Gamage, S. and Watson, I.B. (eds.) *Conflict and community in contemporary Sri Lanka*. Colombo: Vijitha Yapa, pp. 197–216.

Perera, S. (2001) Spirit possessions and avenging ghosts: Stories of supernatural activity as narratives of terror and mechanisms of coping and remembering. In: Das, V., Kleinman, A., Lock, M., Ramphele, R. and Reynolds, P. (eds.) *Remaking a world*. Berkeley: University of California Press, pp. 155–200.

Pupavac, V. (2001) Therapeutic governance: Psycho-social intervention and trauma risk management. *Disasters*, 25(4), pp. 358–372.

Simpson, E. and de Alwis, M. (2008) Remembering natural disaster: Politics and culture of memorials in Gujarat and Sri Lanka. *Anthropology Today*, 24(4), pp. 6–12.

Somasunderam, D. (1998). *Scarred minds: The psychological impact of war on Sri Lankan Tamils*. Colombo: Vijitha Yapa.

Summerfield, D. (1997) Legacy of war: Beyond "trauma" to the social fabric. *Lancet*, 349(9065). Available from doi: 10.1016/S0140-6736(05)61627-3 (accessed November 3, 2015).

Van der Veen, M. and Somasunderam, D. (2006) Responding to the psychosocial impact of the tsunami in a war zone: Experiences from Northern Sri Lanka. *Interventions*, 4(1), pp. 53–57.

9 Japanese families decoupling following the Fukushima Nuclear Plant disaster

Men's choice between economic stability and radiation exposure

Rika Morioka

Sixteen months after the Fukushima Daiichi Nuclear Power Plant exploded, Mrs. Nonaka was still living with a deep fear that they were not safe in Fukushima. Despite her doubts, she could not convince her husband to permanently move away. Sitting on a bench near Fukushima station, where a Geiger counter in a small plot indicated 0.8 microsieverts an hour (700 times higher than in neighboring nations), she lamented that his parents made her cook and eat vegetables grown in their garden, which she suspected were contaminated with radiation. Her husband's insistence on their safety based on the government's claim isolated her in her family. In another household, Mr. Harada, employee of a metal production company, had also made the decision to ignore his wife's warning about the radiation and to stay in Tokyo though his wife, too, repeatedly asked him to move to a safer location. He did not like the fact that the information about the nuclear meltdown had been concealed by the government, but decided to trust the official judgment about the safety of the situation including the level of radiation in drinking water. To him, his responsibility was to be a good breadwinner. His job was too important not only to the financial security of his family but also to his masculine identity to even consider the idea of leaving:

> I want to believe that everything will be fine … I get annoyed when my wife keeps telling me that so and so has moved away and things of that nature because I have no intention of moving away. I am not working for a company that allows me to move. She has asked me why I don't evacuate them, but all I could say was if you wanted to go, you could go. But I will stay here working and making a living.

This chapter examines how the Fukushima nuclear disaster created a social condition in which the traditional notion of breadwinning fatherhood came into conflict with the traditional role of mother as a caretaker, making it harder to protect children from potential radiation exposure. Gender is known to be a key factor in decision-making in disaster settings in intimate relationships. Gender analysis within the household helps to see who has control over important decisions in emergency situations and how power is exercised. Scholars of disaster have noted that the gendered division of labor, particularly women's labor, is known to be a key factor for people to cope with disastrous events (Enarson *et al.*, 2007). The cases presented in this chapter suggest that men's focus on the traditional breadwinning role can produce far-reaching and unintended consequences in a disaster setting.

The division of labor between men and women has been represented as an important factor for the development of Japan, allowing men to focus on economic activities while women attend to domestic responsibilities and serve as a disposable labor force (Houseman and Katharine, 1993; Gordon, 1997). But, when the nuclear disaster in Fukushima put the nation in turmoil, the gender roles that were previously taken for granted created a schism between fathers and mothers. For men, particularly those who worked in powerful institutions, radiation meant a threat to the nation's economic stability and their masculinity embodied in their work and breadwinner roles. For women, the danger of radiation exposure meant threats to their families' health, particularly among children. To explore the implications of this conflict, I first examine why men ignored the radiation risk, considering how masculinity and the traditional division of labor in Japan have historically been linked to the economic interests of the nation state. I then describe the ways in which husbands and wives interpreted and reacted to the radiation crises differently after the explosion, and explain how fathers' responses to the radiation disaster shaped family dynamics and constrained women's ability to act in the aftermath.

The qualitative data presented in this study were gathered mainly through interviews with men and women loosely linked to antiradiation activist networks in Tokyo, Fukushima and Sendai, Japan. Twenty-seven in-depth interviews were conducted between November 2011 and July 2012, using snowball sampling to first contact individuals who were actively seeking protection from radiation after the disaster, and then their spouses who did not. The men and women interviewed were mostly in their 30s and 40s with high school and above education. The author lived in northern Japan for six months and worked for emergency responses during the aftermath of the 2011 earthquake and tsunami disaster, conducting numerous participant observations in informal meetings with concerned parents. In addition, the content of an online forum of a parents' group which had about 210 memberships was analyzed.

Why men ignored the radiation risk

The nation state has been the overpowering frame of reference for Japanese masculinity since the late nineteenth century (Frühstück and Walthall, 2011). Mason (2011) argues that there is a symbiotic relationship between individual and national bodies in Japan: The nation's strength and moral rectitude are articulated through citizens' physical and spiritual inheritance, and unfavorable conditions or weaknesses in individual bodies are seen as a crisis in the national body. Japan's body politic that engaged the male body in its imperialist and militarist rhetoric in the earlier part of the last century eventually shifted to the realm of the economy and businesses. For the second half of the twentieth century, the salaryman (white-collar breadwinner ideally employed by a large corporation) marked the epitome of masculine maturity. The ghost of the salaryman still haunts contemporary Japan in unexpected ways (LeBlanc, 2012), including in the new ideal of technologized masculinity (the *otaku*, or men with obsessive interests, particularly with virtual realities); nostalgia for happy modern marriage (Napier, 2011); and rejection by workers of new masculinity models proposed by labor unions in favor of the dominant salaryman ideal (Gertis, 2011).

The rising level of radiation from the Fukushima plant that endangered the health of the land was often interpreted as a threat to the national body, particularly by those in government and businesses. Beck (1992) argues that physical risks are created and effected in the very social systems designed to manage the risk activity. As Beck posits, the issue of trust is at the core of technological risks in modern societies because physical risks arise from social dependency upon institutions that are often inaccessible to most people affected by the risk.

After the initial evacuation of some Fukushima residents, the government insisted on the harmlessness of the situation even in villages near the damaged power plant (Onishi and Fackler, 2011). The Ministry of Agriculture, Forestry and Fisheries led the effort, for example, by organizing "Eat and Support" fairs promoting local products grown in disaster-stricken northern Japan. Private businesses such as the natural gas company Tosai Gas, which serviced the affected areas, collaborated with the local government to organize similar seafood fairs. Their website banner read "Buy, Eat and Support Recovery!" in order to promote seafood harvested in the area. Japan's single-minded focus on the economic recovery has been termed "disaster nationalism," a matter of concern for many nations (Hornung, 2011). The Japanese government's claims to safety with the support of pro-nuclear scientists and media (*daijyōbu no gasshō*) set the backdrop for the fathers interviewed to downplay the risk of radiation and to seek to resume economic activities.

Studies have consistently shown that health risks posed by technology are often perceived as more acceptable by men than women, including professionals in science (Slovic *et al.*, 1995; Gustafson, 1998). White males with better education, income and conservative views were found to be more trusting of authorities and to have less concern about environmental risks than black men or women (Finucane *et al.*, 2000; Palmer, 2003), suggesting the importance of gender, power and structural factors.

Central institutions are dominated by males, and men therefore trust the judgment of government, science and technology institutions more than women (Fox and Firebaugh, 1992; Flynn *et al.*, 1994). They may also be motivated to protect masculine identities through their commitment to and trust of dominant controlling and regulatory institutions (Bickerstaff, 2004; Kahan *et al.*, 2007). The historically privileged positions of men and their membership in the most advantaged group further socialize them toward risk-taking (Kalof, 2002). I argue below that because their sense of masculinity was deeply rooted in the nation's ability to bring economic stability to their lives, Japanese fathers in dominant institutions consciously chose to trust the government even when they were not satisfied with the latter's handling of the crises.

The nation state at home

Despite becoming a leading industrial society, Japan's social order has remained paternalistic, stressing the fulfillment of ascribed roles (Rohlen, 1974). In mid-twentieth-century economic development, the embrace of pre-industrial traditions was often seen abroad as an exceptional Japanese trait, one allowing for rapid economic growth with minimum social disruption (Goode, 1963). Though less widely known than Japanese management styles, the traditional division of labor between men and women has been one of the main factors supporting Japanese development. The state and corporations actively constructed gender roles to ensure that households were run efficiently by women so that men could provide labor power and devote themselves to economic activities (Gordon, 1997). Women typically entered the labor force before marriage and returned to it as part-time or nonpermanent auxiliary employees at the end of childrearing. In this way, Japanese women have provided a "buffer" labor force in the economy allowing employers to lay off female workers instead of male breadwinners in times of economic downturn (Houseman and Katharine, 1993).

When the traditional gender division of labor met the postwar egalitarian family ideal, gender equality came to be understood as separate spheres of influence for women and men (White, 2002). Fathers and mothers in Japan often live in two separate worlds, with men spending most of their time in political and economic spheres. Patriarchal expectations continue to delegate the caretaker role to women, regardless of their employment status, and the breadwinner role to men, although the realities of family life are now far more diversified than the traditional salaryman family (White, 2002) and women do find socially acceptable ways to voice their political opinions (LeBlanc, 1999; Morioka, 2013).

The reliance of the government on the family as the primary social welfare system also rests on the assumption that it is women's principal responsibility, regardless of their income level, to care for family members, particularly for children. In reality, very few couples manage the division of labor without a stay-at-home mother (White, 2002), resulting in further reinforcing structural gender inequality. Japan consistently ranks among the worst according to conventional gender equality measures. For example, the Gender Gap Report places the country 105th out of 136 countries in its 2013 gender equality index (World Economic Forum, 2013). Furthermore, the United Nations Development Programme's Gender Inequality Index for 2013 reveals that fewer than 10% of parliamentary seats are occupied by women in Japan (UNDP, 2013). In contrast to low levels of female political participation, women perform 87% of household labor (Fuwa, 2004). These patterns prove important in decision-making around post-disaster threat.

Economic dimensions of masculinity

As economic stability and work have been an integral part of Japanese masculinity, work organizations play an important role in men's lives. In the collective efforts toward development and stability, postwar ways of working tended to disregard individual differences and needs, leaving work as the central life interest for men. Full-time workers, by definition, have little control over their time and view long working hours as a demonstration of social integration (Hisamoto, 2003). Free-lance writer Kou Suzuki (2012) posits that Japanese men have learned to conflate their work organizations with society as a whole. In the minds of Japanese men, company and nation are synonymous, and the prosperity of both is closely tied to nuclear energy. Japanese employees thus come to believe that the fate of their companies, the economy and the nation rests on their shoulders, as male bread-winners, thus leading them to rationalize the risks associated with nuclear power despite the ambivalence of many. As LeBlanc noted (2012), debates about nuclear power in Japan have been polarized between the dangers of nuclear power and the risk of economic decline. Behind the debate lurks an unarticulated but framing concern about whether a nuclear-free Japanese economy would still allow bread-winners to feed their families.

Stability in employment and life have historically been an important factor in Japanese modern families (Gordon, 1997; White, 2002), resting in large part on a sense of masculinity that accepts dependence on employer paternalism (Sakurai, 2011). In my own research, the explanations given by Mr. Harada, a corporate employee, illustrated this:

> My wife and son evacuated, but I didn't go because I had my work ... My wife says we can always sell the house, but I value the work I do now. It is because of my work that we have been able to live the way we do ... I need to pay for living expenses, mortgage, and school fees. If I was alone, I think I could make a living anywhere. But I have to think about my child's future. I am not worried about myself ... I am not a university graduate. Because I learnt many things along the way, I am who I am now. Now I came to a point where I can teach my subordinates. Honestly, I think it was this company that made me who I am today ... My work takes a large part of me.

Not all fathers resisted evacuation. Mr. Yamamura, a freelance professional and the father of a two-year-old, let his family evacuate. When his action was not understood by others, he attributed this to the fact that most men worked for *kaisha* (companies) and were "shackled" by work demands and by a corporate culture that emphasized masculine toughness and economic prowess. In his words:

> They just didn't have the sense of alarm ... Men lack imaginations. They don't ask why not evacuate or if it is stupid not to. If death is certain, they would run away. But when I say something, they would say "what about my livelihood, are you going to take care of us?" The media and the government subtly take advantage of this. The men don't generally try to move away even from hot spots [highly irradiated areas], saying I have work to do. Even when things feel like not OK, they prioritize work.

He stressed that being a freelancer made it easier, not harder, to evacuate his family though his income was less stable than those of salarymen. He thought most men working for companies were unable to speak up about their worries or take action, even if they so desired, because of a corporate environment explicitly downplaying the risk. There was a mood in work organizations that rejected the expression of concern or open discussions of the radiation problem. Mr. Harada elaborated the point from his own experience:

> There was a colleague in my office who said to me "It is troublesome that some people are creating maps of radiation levels, asking for the measurement of radiation in milk provided in school lunch, and removing contaminated soils. I don't like it." I think he didn't like it because if something came up, he would have to do something about it.

The concerns expressed by informants in this study demonstrated that the polarized debates about safety, risk and radiation were often gendered between men in powerful institutions and women concerned about health risks. In the disaster-affected areas of northern Japan, people self-censored what to say and refused to discuss the threat of Fukushima radiation in an attempt not to "make matters worse" (Morioka, 2013). It was concern for jobs, industries and the economy that made people stay silent despite the profound sense of alarm expressed by worried mothers. Mr. Kato, who formerly worked in the fishery industry in northern Japan, made this point clearly:

> The easiest thing is to say that you are worrying too much ... If you make a fuss about it, it would make trouble for the people in agriculture and fishery. People say the locals should eat local food. We have to recover from the disaster. Since the government says it is safe, so it must be safe, that's how most people think ... Men around me don't show much interest in the issue. For men, work is the priority. After coming back from work, it is too much to think about it with mothers ... Everyone is tired from work and is feeling like being cornered, and unable to discuss whether it really is safe or not with mothers.

Why women were worried

Mothers of small children, who were somewhat removed from the influence of dominant norms and institutions, often differed in their interpretation of what the crises meant, and particularly about what fathers should be doing. In their view, the disaster meant a threat to the lives of children and families, and fathers were to do everything they could to *protect them* from imminent radiation exposure, putting this first before the health of the economy (Morioka, 2014). This created conflict in the family, as concerned mothers tried to gather information about radiation, learn how to avoid potentially irradiated food and take actions to minimize the contamination. Some participated in mothers' forums to support one another. Yet, at home they were often discouraged, constrained or undermined by fathers' unwillingness to acknowledge the potential danger. Emoto Saori, a Tokyo housewife and mother of a six-year-old explained:

> Rather than radiation, he is thinking about his work. I feel he should be worried more about it. I am trying to gather information about safe food, but he doesn't even try to hear about it. I am discontented. He just does things like buying an ice cream to give to our daughter. We know that milk is not safe, and I try not to buy them since we don't know what's in them ... But he is just not worried about it.

In some cases, fathers not only refused to discuss the issue, but they also made it difficult for mothers to express their concerns for children. Some fathers became angry at mothers' persistent worries. Mr. Kato, a self-employed man who was actively involved in the radiation issue along with women, reflected on this:

> Women are trying hard, but men are pathetic. If men have no time, there are other ways [to help]. I often participate in the forum with mothers, and hear that fathers stand in the way. They don't approve of it. Fathers tell mothers they worry too much and are being *shinkeishitsu* [neurotic]. They should at least make it easier for mothers to take action without interfering. Mothers often ask me how they can convince their husbands, and what they should be doing without their husbands understanding. Mothers with children under elementary school age are worried because their husbands don't understand them.

Concerned mothers found it difficult to hold open discussions about the future of their families or how best to protect children when their husbands became upset and angry for pressing the issue. Their concerns for radiation were often silenced by fathers who did not wish to act on their fear, for instance by evacuating as a family or supporting the evacuation, of their wives and children. A woman in her 40s in Fukushima articulated the profound gulf she saw between mothers and fathers:

> When I said to my husband that I wanted to evacuate my children, he came to ask me why in turn. "Why do you ask why?" I said. It felt like he was saying how come you worry that much. There is an unspoken understanding that fathers must work while they are raising children, and that's why he prioritizes economic development. They should think for themselves about what the government is doing. They just follow what the top is saying. I sense the brittleness of my husband being unable to prepare and face the difficult realities ... Mothers can even divorce their husbands if it is for the sake of protecting children. My friend did.

Not only fathers but other male authority figures derided their concerns, as Kodama Yuko, a Sendai resident and mother of a four-year-old recounted:

They [men] believe it is safe. If I ask questions, it feels like they are saying "how dare you defy the government" (*okuni ni sakarautoha nanigotozo*).
Even my doctor said to me "What in the world are you saying?" (*nani itten
no?*) I want to find a doctor who would check my child properly, but can't find
one. They have already made up their minds that things are fine, so I stopped
asking questions … Even my husband, when I tell him about people who have
evacuated, he would start getting angry saying "what do you want me to do?"

The position of Kanno Norio, the village head of Iidate, one of the most heavily
irradiated areas in Fukushima Prefecture, epitomized the attitude of authorities
with which these women found it difficult to agree. In his book describing the
disaster, Mr. Kanno expressed his frustration at having to evacuate his village
without regard to "stability of life" (Kanno, 2011, p.151):

In the situation that does not allow us to foresee what might be ahead, I
want you to think about what will happen to the six thousand villagers. They
will have to bear mental, physical, and economic risks, even for children's
education. Protecting health from radiation is heavily regarded as the most
important, and the risk of destroying the stability of life is not taken into consideration. Even though there are scholars who would argue that the risk of
developing cancer from smoking is higher than that of radiation.

Those who were concerned with the health consequences of the disaster were
frustrated with the slow response of the government. Many engaged in activism
to try and change the situation (Morioka, 2013; Slater *et al.*, 2014), sometimes
in the face of clear opposition as was the experience of Sendai housewife Keiko:

I feel the pressure of the government. I don't watch TV anymore. I look at the
internet. I was gathering information, and found out about the [antinuclear]
demonstration in Tokyo. When I told my father about it, he yelled at me saying "how can there be such a thing?"

Another example of contrasting reactions between men and women was cap-
tured in a video of a 2012 hearing held by the Nuclear and Industrial Safety
Agency (NISA) to discuss the restarting of Kansai Electric's Ooi Nuclear Power
Plant (tokyobrowntobby2, 2012). Upset both with the prospect of the reopening of
the nuclear plant and the decision by NISA to lock them out of the room, a female
group of protesters came into the meeting room with a TV camera. A woman from
Fukushima emerged from the crowd and asked the group of male experts whether
they were "man enough" to protect women and children. The impassive male
experts in front did not answer her question, but tried to continue with the meeting by stressing the presumed economic benefits of reopening the nuclear power
plant. The emotional plea of the women and the business-like attitude of the stern
male experts were in clear contrast. One veteran antinuclear activist, Ms. Saito in
Fukushima, explained her thoughts this way:

There are huge differences in values between men and women. Lots of "disaster divorces" are occurring too because of that. There are still so many men who can't take off business suits. They keep saying "what will happen to our economy?" We don't have to live like the rich. We don't have to pay for loans. If our income was one thousand a month, we could just live with that.

The contrasting perspectives of mothers and fathers described above, with women expecting men to protect families' health at all costs, including by taking economic risks, help explain why "disaster divorces" were a common outcome of the disaster (Miura, 2012). Studies have shown that conflicting interests of male and female members of the family unit often result in changes in household composition during and after disaster. One of the consequences of this, particularly for women and children, is increased vulnerability and weakened capacity to cope with future disaster (Wiest, 1998).

Conclusion

Extreme events such as a nuclear disaster make obvious what is normally hidden in day-to-day practices (Oliver-Smith and Hoffman, 1999). In the case of Japan, the radiation crises in the aftermath of the Fukushima explosion exposed the gender schism that was hidden under the guise of the traditional division of labor between men and women. Spouses found it hard to reconcile their differences when the crises forced them to make urgent decisions though faced with uncertainty about critical issues that encompassed both men's and women's spheres. They were unable to openly discuss and make consensual decisions about how best to protect their children from the imminent threat of radiation. The term *bundan* (decoupling) was often used to describe this social schism. Disjuncture that married couples managed to disregard in their hectic day-to-day lives before the disaster were later manifested openly as *bundan*, visible not only in families but also in communities and the nation. What was at the bottom of these divided families was the separate and detached interests of fathers and mothers, born in turn of their separate and gendered worlds as Japanese men and women.

To be sure, divisions within the nation did not only appear along gender lines. Many men opposed nuclear energy and many women agreed with government and corporate policies. Yet the gender gap was significant, pointing to masculine norms as key drivers of the nation's responses to a major nuclear disaster. The unwillingness of the local government to measure radiation levels, the immediate post-disaster increases in allowable maximum radiation limits in food, and the promotion of potentially harmful local agricultural and fishery products by the authorities were each examples of public policies reflecting the values of a government overwhelmingly run by men and of fathers such as those interviewed in this study. Like fathers, the government sought to protect hyper-masculinized notions of economic recovery and national stability, primarily by reinforcing traditional notions of masculinity and the gender division of labor in private and public life. In contrast, women who were dissatisfied with governmental inaction prioritized the protection of vulnerable life.

These divisions and contrasts represent critical forces to be examined in contexts of national disruption, for they illumine the depth of resistance to change that may exist before a disaster at both the family and institutional levels. The finding above reported that the small numbers of men who protested with mothers tended to be outside the ranks of dominant institutions (e.g. freelance professionals, retirees, students). This may provide an entry point for inquiries about the influence of powerful institutions and the larger social and organizational environment on gendered risk perceptions and recovery processes. Further analyses are critically needed to better take into account how gender and dominant notions of masculinity influence the social context in which disaster policies and tolerance of risk develop.

References

Beck, U. (1992) *Risk society: Towards a new modernity*. New York: Sage.

Bickerstaff, K. (2004) Risk perception research: Socio-cultural perspectives on the public experience of air pollution. *Environment International*, 30(6), pp. 827–840.

Enarson, E., Fothergill, A. and Peek, L. (2007) Gender and disaster: Foundations and directions. In: Rodriguez, H., Quarantelli, E. and Dynes, R. (eds.) *Handbook of disaster research*. New York: Springer, pp. 130–146.

Finucane, M., Slovic, P., Mertz, C.K., Flynn, J. and Satterfield, T. (2000) Gender, race, and perceived risk: The "white male" effect. *Healthy Risk & Society*, 2(2), pp. 159–172.

Flynn, J., Slovic, P. and Mertz, K. (1994) Gender, race, and perception of environmental health risks. *Risk Analysis*, 14(6), pp. 1101–1108.

Fox, F. and Firebaugh, G. (1992) Confidence in science: The gender gap. *Social Science Quarterly*, 73(1), pp. 101–113.

Frühstück, S. and Walthall, A. (2011) *Recreating Japanese men*. Berkeley: University of California Press.

Fuwa, M. (2004) Macro-level gender inequality and the division of household labor in 22 countries. *American Sociological Review*, 69(6), pp. 751–767.

Gertis, C. (2011) Losing the union man: Class and gender in the postwar labor movements. In: Frühstück, S. and Walthall, A. (eds.) *Recreating Japanese men*. Berkeley: University of California Press, pp. 135–153.

Goode, W. (1963) *World revolution and family patterns*. New York: The Free Press.

Gordon, A. (1997) Managing the Japanese household: The new life movement in postwar Japan. *Social Politics*, 4(2): pp. 245–283.

Gustafson, E. (1998) Gender differences in risk perception: Theoretical and methodological perspectives. *Risk Analysis*, 18(6), pp. 805–811.

Hisamoto, N. (2003) *Seishain runesansu* [Regular employee renaissance] Tokyo: Chuko Shinsho.

Hornung, W. (2011) The risks of "disaster nationalism." *Pacific Forum CSIS*, Number 34. Available from http://csis.org/publication/pacnet-34-risks-disaster-nationalism (accessed December 12, 2012).

Houseman, N. and Katharine A. (1993) Female workers as a buffer in the Japanese economy. *The American Economic Review*, 83(2), pp. 45–51.

Kahan, D., Braman, D., Gastil, J., Slovic, P. and Mertz, C. (2007) Culture and identity-protective cognition: Explaining the white male effect in risk perception. *Journal of Empirical Legal Studies*, 4(3), pp. 465–505.

Kalof, L., Dietz, T., Guagnano, G., and Stern, P.C. (2002) Race, gender and environmentalism: The atypical values and beliefs of white men. *Race, Gender & Class*, 9(2), pp. 112–130.

Kanno, N. (2011) *Utsukushiimurani houshasenga hutta* [Radiation rained on a beautiful village]. Tokyo: Wani Books.

LeBlanc, R. (1999) *Bicycle citizens: The political world of the Japanese housewife*. Berkeley: University of California Press.

LeBlanc, R. (2012) Lesson from the ghost of salaryman past: The global costs of the bread-winner imaginary. *The Journal of Asian Studies* 71, pp. 857–871. Available from doi: 10.1017/S0021911812001209 (accessed November 9, 2015).

Mason, M. (2011) Empowering the would-be warrior: Bushido and the gendered bodies of the Japanese nation. In: Frühstück, S. and Walthall, A. (eds.) *Recreating Japanese men*. Berkeley: University of California Press, pp. 68–90.

Miura, A. (2012) *Shinsai rikon* [Disaster divorce]. Tokyo: East Press.

Morioka, R. (2013) Mother courage: Women as activists between a passive populace and a paralyzed government. In: Gill, T., Slater, D. and Steger, B. (eds.) *Japan copes with calamity: Ethnographies of the earthquake, tsunami and nuclear disasters of March 2011*. Oxford: Peter Lang, pp. 177–200.

Morioka, R. (2014) Gender difference in the health risk perception of radiation from Fukushima in Japan: The role of hegemonic masculinity. *Social Science & Medicine*, 107, pp. 105–112.

Napier, S. (2011) Where have all the salarymen gone? Masculinity, masochism, and tech-nomobility in *densha otoko*. In: Frühstück, S. and Walthall, A. (eds.) *Recreating Japanese men*. Berkeley: University of California Press, pp. 54–176.

Oliver-Smith, A. and Hoffman, S.M. (eds.) (1999) *The Angry earth: Disaster in anthropological perspective*. New York: Routledge.

Onishi, N. and Fackler, M. (2011) Japan held nuclear data, leaving evacuees in peril. *The New York Times*. August 8. Available from http://www.nytimes.com/2011/08/09/world/asia/09japan.html?pagewanted=all. (accessed August 10, 2011]).

Palmer, C. (2003) Risk perception: Another look at the "white male" effect. *Health, Risk & Society*, 5(1), pp. 71–83.

Rohlen, T.P. (1974) *For harmony and strength: Japanese white-collar organization in anthropological perspective*. Berkeley: University of California Press.

Sakurai, Y. (2011) Perpetual dependency: The life course of male workers in a merchant house. In: Frühstück, S. and Walthall, A. (eds.) *Recreating Japanese men*. Berkeley: University of California Press, pp. 115–134.

Slater, H., Morioka, R. and Danzuka, H. (2014) Micro-politics of radiation: Young mothers looking for a choice in post-3.11 Fukushima. *Critical Asian Studies*, 46(3), pp. 485–508.

Slovic, P., Malmfors, T., Krewski, D., Mertz, C.K., Neil, N. and Bartlett, S. (1995) Intuitive toxicology II. Expert and lay judgments of chemical risks in Canada. *Risk Analysis*,15, pp. 661–675.

Suzuki, K. (2012) Otoko to onna to gennpatsuto [Men, women and nuclear energy]. *Tokidoki Osanpo Nikki 86*, Magazine 9. Available from http://www.magazine9.jp/osanpo/120321 (accessed December 4, 2012).

tokyobrowntobby2 (2012) Stress test meeting interrupted: Audience's protest and poignant words from a Fukushima woman. *YouTube*. Available from https://www.youtube.com/watch?feature=player_embedded&v=hLYrZsCQsko (accessed December 6, 2012).

UNDP (2013) Human development reports, Gender inequality index. Available from http://hdr.undp.org/en/content/table-4-gender-inequality-index (accessed August 27, 2015).

White, M. (2002) *Perfectly Japanese: Making families in an era of upheaval*. Berkeley: University of California Press.

Wiest, R. (1998) A comparative perspective on household, gender, and kinship in relation to disaster. In: Enarson, E. and Morrow, B. (eds.) *The gendered terrain of disaster: Through women's eyes*. Westport, CN: Praeger, pp. 63–79.

World Economic Forum (2013) *Gender gap report*. Available from http://www.weforum.org/reports/global-gender-gap-report-2013 (accessed August 27, 2015).

Part III

Diversity of impact and response among men in the aftermath of disaster

10 Disabled masculinities and disasters

Mark Sherry

Disabled men's subjectivities are uniquely shaped by abjection. In disasters, the results of such abjection are life-threatening. In every disaster in the last 20 years, disabled people died and were injured at higher rates than nondisabled people because of inaccessible environments. As a result of such disabling barriers, abjection and the power of normativity, disabled masculinities experience unique social expressions and social locations—even though they are shaped by other forms of power such as race, class and gender. Disabled masculinities are interstitial and indeterminate—a position that offers possibilities for both "gender trouble" and "normativity trouble" which can challenge hegemonic masculinity in multiple contexts, including disasters.

Disabling barriers and disasters

In every disaster in the last 20 years, a lack of accessibility has consistently resulted in lower rates of evacuation, higher rates of injury and higher mortality rates for disabled people. For examples of this phenomena, see case studies of the 1995 Hanshin earthquake, the 2001 terrorist attacks on the World Trade Center, the 2004 Asian tsunami, 2005 Hurricane Katrina, 2007 Cyclone Sidr in Bangladesh, the 2009 Southern California wildfires, the Christchurch earthquakes of 2010–2011, the 2011 Japanese earthquake and many others (Rahman and Mallick, 2007; Takahashi *et al.*, 1997; Davis and Phillips, 2009; Nakamura, 2009; Liu *et al.*, 2012; Phibbs *et al.*, 2015). Warnings, transportation, medical care, evacuation and sheltering are consistently inaccessible. Barriers in disaster support, lack of specialized disability-related supports and medical supplies, failure to provide assistive technology and equipment, lack of adequate rehabilitation services, and failure to engage in inclusive recovery also result in countless avoidable deaths. Inaccessible communication systems, temporary refuges and evacuation systems cost disabled people their lives.

The disability rights movement is one of the most important influences on the subjectivities of disabled men. By linking disability to social, political, cultural and economic subjugation, the movement shifted disability from the private to the public realm. For many disabled men, repositioning disability from a personal tragedy to a focus on profound social devaluation lies at the heart of their identity, political commitment and ethics. In the context of the harmful effects of inaccessible environments, disabled masculinities often reflect the influence of a "social model of disability" (Oliver, 1990, p. 1) which positions disability not as a personal tragedy but a form of oppression. The social model inspired disabled people to consider such inaccessibility a public, political issue. Applying the social model to disaster suggests that disaster response, preparation and emergency procedures are essentially designed for nondisabled people. Nondisabled privilege has resulted in disabling barriers in disaster planning and support, lack of specialized disability-related supports, failure to provide assistive technology and rehabilitation services, failure to engage in inclusive recovery and much more.

Statistics hint at (but do not do justice to) the magnitude of the injustice inflicted on disabled people in disasters because of inaccessibility. The 2011 Japanese earthquake and tsunami provided clear evidence of disparities in mortality: people with physical, intellectual and psychiatric impairments were 2.32 times more likely to die than their nondisabled peers (Liu *et al.*, 2012). In Miyagi Prefecture, hearing impaired people experienced the highest mortality rate because warnings were given through an inaccessible communication system—sirens (Gonon, 2013). There were also disproportionate deaths of disabled people in the 2004 Asian tsunami. For instance, in Port Blair, South Andaman, there were 700 people with polio and none survived, and in the post-tsunami recovery efforts in India, food and water was distributed on a "first-come, first-served" basis, systematically disadvantaging those with mobility impairments (Singh, 2007).

The material circumstances of disabled people, which are far worse than those experienced by nondisabled people, shape their life course in a myriad of ways. For example, the systemic impoverishment of disabled people greatly influences their chances of surviving a disaster. Because 82% of disabled people in the Global South live in poverty (Parnes *et al.*, 2009), they are more likely to live in disaster zones (Carrigan, 2010). As well, they are often unable to afford the types of disaster preparedness resources which wealthier people can afford, increasing the chances that they will die.

Disproportionate deaths also occur in the Global North. In the Hurricane Katrina evacuations, the lives of some disabled people were not given the same priority as nondisabled people. At Memorial Hospital in New Orleans, people with the most severe impairments (particularly those on ventilators) were killed by lethal injections (given without their knowledge or consent) instead of being evacuated (Fink, 2013). Multiple testimonies from disabled people over the years are a powerful reflection of the deathly (and sometimes eugenic) disablism which occurs during disasters—unfortunately, space prevents them from being included in this chapter.

Another barrier is that staff in humanitarian agencies are often unaware of the specific needs of disabled people in emergency contexts and they typically focus on newly impaired people, ignoring the needs of people who were already disabled (Kett and van Ommeren, 2009). They may also be reluctant to change their practices in order to be more inclusive. One US study found that two-thirds of disaster responders refused to modify guidelines to include disability accommodations (Fox *et al.*, 2007). Public health concerns may also be deployed to exclude disabled men from the resources of the nation, particularly if they are from another culture. In such cases, racism blends with disablism to form a rubric of exclusion and prejudice.

Even though a great deal of literature recognizes that disabled people consistently experience worse outcomes in disasters, the existing literature has been limited by a narrow approach which assumes disability is a medical issue, or a narrow social model approach which simply identifies specific disabling barriers and demands their removal. What is lacking is a broader analysis of normativity as a wider system of power. Normativity is particularly important when discussing disabled masculinities.

Disabled masculinities, hegemonic masculinity and normativity

The term "disabled masculinities" is a theoretical concept still under development, but it seems to have multiple meanings. It can refer to: performative enactments involved in the process of doing gender; regulatory ideals; hegemonic or alternative discourses, traits and/or behaviors; a particular identification understood within the wide arrange of sex/gender possibilities; a form of subjectivity; and/or patterns of systematic privilege or disadvantage for particular groups of disabled men (or disabled men, collectively, over disabled women). Disabled masculinities are therefore always complex, contested and multiply situated— whether they are acquired or genetic, their meanings and significance are socially constructed. They are also culturally encoded with strong messages about life and death, loss and grief, equality and dignity, respect and recognition, and human rights and survival. And they are indelibly meshed in the social construction of hegemonic masculinity and normativity. These vectors of power play out in both casual and formal settings, in institutional policies and in extraordinary moments for the body politic, such as those which occur in disasters.

Like hegemonic masculinity, normativity looms large in terms of disabled masculinities. Disability is always framed in contrast to socially constructed assumptions about normality. Normativity therefore produces multiple variations of disabled bodies, identifications, representation, ethics and behavior. In the same way as hegemonic masculinity operates as a form of power, normativity helps inform disabled men who they are, who they should aspire to be, what sort of body they should aspire to have, what behavior might be more or less socially valued and so on. Normativity always tends to reproduce hegemonic forms of inequality and domination. In such a context, disabled masculinities are presented as flawed or incomplete. But studies of various impairments have suggested that the experience of disabled masculinities is far more complex. For instance, one study of masculinity and intellectual disability suggested that the experiences of these men were so complex that "a potential reformulation of gendered notions of dependence, marginalization, powerlessness, and disadvantage" was required (Wilson *et al.*, 2012, p. 262).

However, normativity is never completely successful—disabled masculinities are also a site of resistance and innovation. Affective relationships between bodies are far more complex than a simple disavowal of disability; possibilities abound for alternative interpretations. Disability always has the potential to contest hegemonic discourses, produce alternative aesthetics and subjectivities, and challenge normative expectations (Siebers, 2010). Disabled masculinities therefore raise the possibilities for "rethinking the terms of ability and gender domination" (Torrell, 2013, p. 210) and expanding "the masculine repertoire" by developing flexible embodied sex and gender roles (Shuttleworth, 2004). Indeed, the ways in which disabled masculinities often refuse those stereotypes of men as strong risk-takers who put their bodies on the line may well be a lesson for nondisabled men, who die in disasters because of such risky endeavors (Tyler and Fairbrother, 2013).

When impairments are newly acquired, disabled masculinities are multiple and fluid, because there are multiple ways of making sense of new forms of embodiment, new identities and new lifestyles. One option available to disabled men is to adjust and rework their notions of masculinity. They may abandon some of the hegemonic notions of masculinity such as demonstrating stoicism and refusing to discuss medical issues, replacing them with alternative attitudes and behaviors. Such alternative masculinities can include an emphasis on interdependence rather than independence, increased openness about personal frailty and health issues, and a greater willingness to engage in self-disclosure. New forms of emotional expression may also become available, opening their range of masculine repertoires to include possibilities associated with notions of vulnerability and simultaneously providing/receiving care.

The apparent contradictions between disabled masculinities and hegemonic masculinity cannot be denied. Hegemonic masculinity emphasizes self-reliance, refuses care, valorizes physical activity, devalues a lack of mobility, and discourages discussions of pain (unless such discussions are framed in the context of overcoming it, or pushing through it). Disabled men are necessarily marginalized and devalued because of their assumed inability to meet these norms. Indeed, some men with physical impairments are labeled "medically fragile" (Ng *et al.*, 2015) and experience "medicalized masculinities" (Rosenfield and Faircloth, 2006), far removed from nondisabled hegemonic norms. On the other hand, men with intellectual impairments are often positioned as "conditionally masculine"— their experiences of masculinity are ambiguously framed at the intersection of social, embodied and cognitive domains. Nondisabled privilege has meant that cognitive considerations are frequently omitted from discussions of masculinities. Regardless, they deeply influence the impact of normative gender expectations. For men with intellectual disabilities, support staff may have unique influence on the expression of gender, masculinity and sexuality (Gill, 2015).

While normativity (by definition) is nondisabled, some disabled men are still enticed to attempt its impossible ideals. The unending quest by these men for sexual, gendered and bodily identities or behaviors which approximate normativity demonstrates its power as a regulatory ideal. The promise of disabled masculinities somehow matching hegemonic norms dangles like a carrot for many men. Their engagement with the disability movement is often discursively influenced by (if not wholly embracing) hegemonic masculinity. For instance, disability advocacy is commonly framed in military terms, such as a "fight" for equality and a "battle" for equal rights (Vaughn, 2003).

But instead of disability being invoked as a metaphorical site for the anxieties and fears of a nation in crisis, disabled masculinities that utilize a disability rights discourse may also involve alternative subjectivities which are built on notions of equality, recognition and inclusion. By challenging disabling barriers and pointing to alternative possibilities for societal reorganization, these masculinities open a space for the renegotiation and recognition of more egalitarian futures and alternative masculine subjectivities. Potentially, normativity and hegemonic masculinity may be open to contestation when the association of disability with alterity, disavowal and abjection collides with discourses of disability that are thoroughly infused with notions of equality, acceptance and inclusion.

Embodiment affects masculinity and masculinity affects embodiment. For men who develop physical impairments as the result of a disaster, rehabilitation is intimately connected with the reconstruction of new disabled subjectivities. Throughout rehabilitation, hegemonic masculinity is constantly invoked. They too are urged to become involved in a "fight," "battle" or "war" against their impairments (Gerber, 2001). To "fight" against a disease or impairment requires traits associated with hegemonic masculinity, such as determination and a relentless struggle to overcome pain and trauma. Some disabled men valorize those forms of hegemonic masculinity which they can still approximate, such as strength and resilience, stoicism, self-reliance or sexual performance.

War and war-related disasters often lead men to develop new impairments, such as Traumatic Brain Injury, Posttraumatic Stress Disorder, amputations and blindness. And since war "is among the most consistently gendered of human activities" (Goldstein, 2004, p. 107), it necessarily creates new expressions of masculinities. Newly disabled veterans typically experience marked changes in their lives and identities after acquiring an impairment. For disabled veterans, the experience of rehabilitation commonly reinforces hegemonic notions of masculinity by linking manliness with being an injured warrior and someone who has paid a price for being a guardian of the nation. Making a sacrifice for a country is often positioned as the ultimate "manly act," which deserves respect and recognition. But such discourses are not always helpful. They may simultaneously position the disabled man as a tragic victim, a form of subjectivity which many disabled men find deeply problematic (Oliver, 1996).

One of the most common effects of disasters is a significant increase in Posttraumatic Stress Disorder (PTSD), which has flow-on effects on newly disabled masculinities. Studies of the 2002 terrorist bombing in Bali, the 2011 massacre in Norway, the 1995 Oklahoma bombing, the September 11, 2001 attacks in the US and many other incidents of terrorism document significant levels of PTSD (Stevens *et al.*, 2013; Neria and Shultz, 2012; Dyb *et al.*, 2014). However, men who are first responders in such incidents, for instance police or firefighters, are members of organizations that are strongly shaped by traditional gender norms. They are enculturated not to express emotions, making it much harder to find safe, supportive spaces to discuss and cope with such experiences. Painful and traumatic memories are generally not shared among the "brotherhood" of first responders (Pasciak and Kelley, 2013). Such cultures of silence are profoundly disabling for those with PTSD.

While the social model of disability prioritizes social over medical concerns, it also seems to be implicitly informed by a disability hierarchy where those with psychiatric, cognitive or intellectual impairments, or those whose impairments are profoundly shaped by trauma and pain, are somewhat marginalized. It would be a mistake to assume that discussions of conditions such as PTSD are a simple reversion to a "medical model" approach to disability. Instead, it is more productive to view it as an engagement with an alternative approach to disability, more akin to the Nordic model, which explores the relationship between the individual and the environment (Mallett and Runswick-Cole, 2014). Such an approach might result in the inclusion of medical conditions which are usually not included in discussions of disability, such as environmental illness.

The engagement of disabled men, even newly disabled men, with the norms of hegemonic masculinity is not a simple one-way process of absolute acceptance. Impairments can shape multiple emotional and embodied responses which render hegemonic masculinity less relevant, if not obviously impossible to achieve. In such situations, interstitial or hybrid disabled masculinities may emerge. These in-between moments suggest "gender trouble" (Butler, 1990) but also what might be called *normativity trouble*. That is, disasters can produce multiple, incoherent and contradictory expressions of gender and multiple challenges to the power of normativity. Simplistic binaries (such as disabled/nondisabled, masculinity/femininity) can become undone, to some degree, when disasters suddenly wreak havoc on a community's sense of normalcy. Hegemonic masculinity seems less dominant when the widespread expression of emotional devastation, loss and grief is commonplace. But if such expression is limited, there is a potential for hegemonic masculinity to conversely be reinforced, limiting the emotional recovery of those who have experienced disasters.

In this context, it might be useful to consider Butler's work on gender as a site of performativity (Butler, 1990). Butler argues that people incessantly improvise gender, within certain constraints, even though they aren't always conscious or deliberative about gender. Navigating disabled masculinities involves "active and tactical patterns of identity construction that implicate a variety of social norms, resources, relationships and contexts" (Barrett, 2014, p. 43). Positioning disabled masculinities as performative implies they may reproduce, reformulate or resist hegemonic norms. Butler argues that sometimes the need to survive leads people to distance themselves from normative standards—and the issue of survival is obviously crucial for disabled people in disasters, given the additional disabling barriers they experience. Such resistance is not simply an individual willful "performance"—Butler argues that it inevitably draws on a collective minority approach. The disability movement has also developed a minority approach in order to address the unique concerns of disabled people.

The unique nature of disabled masculinities

While disabled masculinities share much in common with other masculinities, they are unique in the ways they are indelibly shaped by abjection. Failing to explore the unique elements of disabled masculinities unconsciously removes the need to deconstruct the ways in which wider systems of power and privilege are constituted by abjection. Nondisabled subjectivity is shaped by its relationship of power and privilege to disabled abjection—its disavowal seemingly avoids the need to even discuss the significance of disabling barriers in multiple contexts, including disasters. Disability is not a "special" category relative to nondisability; it is its constituent other, always in a dialogical, iterative and dialectal relation. It is continually remolded with regard to dominant power relations, including nondisability, hegemonic masculinity and normativity. These dominant power relations demand careful examination because of their harmful effects in multiple contexts, including disasters.

One manifestation of these power dynamics is the relegation of disability issues in disasters to the role of "special needs." The discourse of "special needs" is a power-laden form of exclusion that implies disability is a concern for a (presumably needy) minority, irrelevant to nondisabled people—even though they revolve around universal concerns about equality, access and inclusion for everyone. The "special needs" discourse fails to recognize that disabled people could be seen as a litmus test for universal accessibility; if a disaster relief system fails disabled people, it is likely to fail many others too. When disability issues are relegated to "special" needs, the opportunity to explore better solutions for everyone in disasters is missed.

Embodiment is a constantly changing dynamic across the life course, inevitably altered by the processes of aging, individual choices and experiences to various forms of social inequality, including disabling barriers. In contrast to simplistic assumptions about the "master status," permanency and intractability of disability, it is actually an indeterminate embodied location. One environment might disable an individual when another does not; health and impairment can fluctuate; and people may feel more or less disabled in different social locations and contexts, including disasters. Potentially, the unique malleability and indeterminacy of disabled masculinities can interrupt and undermine certain master narratives about the nation, the body, the body politic and hegemonic masculinity—an important dynamic in times of change and disaster.

In different situations and moments associated with a disaster, disabled masculinities may be a source of protection or vulnerability, they may engage in assimilation or transgression, and they may contest or reinforce hegemonic and normative forms of power. "Ablenationalism," the valorization of nondisabled bodies in the collective vision of the nation (Snyder and Mitchell, 2010), inevitably produces resistance. Disabled masculinities resist such ideologies, and in doing so inevitably challenge oppressive social norms and offer alternatives that are far more inclusive. Such gender trouble and normativity trouble highlight the progressive capabilities of disabled masculinities.

Disasters evoke multiple expressions of disabled masculinities. These can vary according to such factors as whether men were already disabled, the ways they acquired their impairments, whether they underwent rehabilitation, whether they were veterans, the type of disaster which they experienced, and their engagement with a discourse of disability rights. The ways in which disabled men "do gender" are also deeply connected to social locations and identities such as race, class, gender, national identity, geographic location and the role of their country in international political economy. For instance, those who become disabled in the Global South are likely to have far fewer resources, support and options than those in the Global North. These layers of power may frame the ways in which a disaster is popularly understood.

Differences among various groups of disabled men are also incredibly important. Disabled masculinities are not homogeneous. Particular groups of disabled men experience privilege in relation to each other, as well as in relationship to women. All men, including disabled men, also experience privilege with respect to the incidence of rape and sexual violence, in everyday situations, disasters and in wars (Zabeida, 2010; Sherry, 2010). Like other men, disabled men's subjectivities should be understood within the wider context of gender inequality. Disabled men are privileged in relation to disabled women with regard to many experiences such as employment opportunities, income and wealth, literacy, experiences of sexual abuse in the life course and mortality rates (Mallett and Runswick-Cole, 2014). In comparison to disabled men, disabled women have been even more neglected in disaster planning and responses, subjected to gender stereotyping that stresses "protection" rather than inclusion and empowerment, and their sex-specific needs have been largely ignored (Enarson, 2009). Control of reproductive rights has also been largely targeted at disabled women through practices such as forced sterilization, forced contraception, forced abortion and forced marriage (Frohmader and Ortoleva, 2013).

However, disabled men's experience of privilege over disabled women does not mean that disabled masculinities simply mirror those of nondisabled men. To ignore or erase the specificity of disabled masculinities is to engage in disablist assumptions about sexuality, gender and normativity. Disabled men's subjectivities are uniquely shaped by socially constructed assumptions about normalcy and embodiment which influence every aspect of their masculinities (Ginsburg and Rapp, 2013). They are framed by a nondisabled medical gaze and wider forms of governmentality that (re)produce particular connections between disabled masculinities and disablism. Such governmentality involves a particular form of epistemological, ontological and experiential invalidation known as "psycho-emotional disablism" (Reeve, 2008, p. 97). To ignore these vectors of power is to reinforce wider social dynamics that situate disability in a marginalized corner; it is thoroughly disablist.

Conclusion

Disabled masculinities are uniquely shaped by abjection, normativity and disabling environments. These power dynamics not only limit the rights of disabled people, and result in increased death rates in situations such as disasters, but they are the largely unrecognized foundation of nondisabled privilege. Unfortunately, the uniqueness of disabled masculinities is not generally recognized—and this absence of recognition is itself a form of disablism, denying the need for changes in the identities and behaviors of nondisabled people. As a result, disabled subjectivities should not be discussed in isolation from nondisabled ones, since they always exist in a dialogic relationship with nondisabled privilege.

Because of abjection, disabling barriers and nondisabled privilege, disabled people consistently experience worse outcomes in disasters compared to nondisabled people. From the planning stage to reconstruction, disabling barriers have been omnipresent in the disasters which were discussed in this chapter. As a result, disabled people have little option but to advocate for improved access and inclusion—and doing gender is an important element of this process. Negotiating access may reproduce elements of hegemonic masculinity. However, it also has the potential to generate new forms of masculinity, freed from the restrictions of hegemonic masculinity and normativity. The solutions to the marginalization of disabled people—the removal of disabling barriers and the provision of universal access—will generate new subjectivities and masculinities. Future research must recognize that the experiences and identities of disabled people, including in disasters, are not a "special" issue that concern a minority of people. They revolve around universal issues of equality, recognition, rights and respect.

References

Barrett, T. (2014) Disabled masculinities: A review and suggestions for further research. *Masculinities & Social Change*, 3, pp. 36–61.

Butler, J. (1990) *Gender trouble: Feminism and the subversion of identity*. New York: Routledge.

Carrigan, A. (2010) Postcolonial disaster, pacific nuclearization, and disabling environments. *Journal of Literary & Cultural Disability Studies*, 4, pp. 255–272.

Davis, E. and Phillips, B. (2009) Effective emergency management: Making improvements for communities and people with disabilities. Washington, DC: National Council On Disability.

Dyb, G., Jensen, T.K., Nygaard, E., Ekeberg, Ø., Diseth, T.H., Wentzel-Larsen, T. and Thoresen, S. (2014) Post-traumatic stress reactions in survivors of the 2011 massacre on Utøya Island, Norway. *The British Journal of Psychiatry*, 204, pp. 361–367.

Enarson, E. (2009) Women, gender & disaster: Abilities and disabilities. Gender and Disaster Network. Available from https://www.Gdnonline.Org/Resources/Gdn_Gender-note4_Abilities.Pdf (accessed October 4, 2015).

Fink, S. (2013) *Five days at Memorial: Life and death in a storm-ravaged hospital*. New York: Crown Publishers.

Fox, M.H., White, G.W., Rooney, C. and Rowland, J.L. (2007) Disaster preparedness and response for persons with mobility impairments: Results from the University of Kansas Nobody Left Behind Study. *Journal of Disability Policy Studies*, 17, pp. 196–205.

Frohmader, C. and Ortoleva, S. (2013) The sexual and reproductive rights of women and girls with disabilities. Issues paper presented at the International Conference on Human

Rights, July 1. The Hague: Netherlands. Available from http://www.wwda.org.au/issues_ paper_srr_women_and_girls_with_disabilities_final.pdf (accessed February 16, 2016).

Gerber, D.A. (2001) Blind and enlightened: The contested origins of the egalitarian politics of the Blinded Veterans Association. In: Longmore, P.K. and Umansky, L. (eds.) *The new disability history: American perspectives*. New York: New York University Press, pp. 313–334.

Gill, M. (2015) *Already doing it: Intellectual disability and sexual agency*. Minneapolis, MN: University of Minnesota Press.

Ginsburg, F. and Rapp, R. (2013) Disability worlds. *Annual Review of Anthropology*, 42, pp. 53–68.

Goldstein, J. (2004) War and gender. In: Ember, C. and Ember, M. (eds.) *Encyclopedia of sex and gender*. New York: Springer, pp. 107–116.

Gonon, A. (2013) Vulnerability in times of disaster. *Iride*, 26, pp. 551–563.

Kett, M. and van Ommeren, M. (2009) Disability, conflict, and emergencies. *The Lancet*, 374, pp. 1801–1803.

Liu, M., Kohzuki, M., Hamamura, A., Ishikawa, M., Saitoh, M., Kurihara, M., Handa, K., Nakamura, H., Fukaura, J. and Kimura, R. (2012) How did rehabilitation professionals act when faced with the Great East Japan Earthquake and disaster? Descriptive epidemiology of disability and an interim report of the relief activities of the ten rehabilitation-related organizations. *Journal of Rehabilitation Medicine*, 44, pp. 421–428.

Mallett, R. and Runswick-Cole, K. (2014) *Approaching disability: Critical issues and perspectives*. London: Routledge.

Nakamura, K. (2009) Disability, destitution, and disaster: Surviving the 1995 Great Hanshin earthquake in Japan. *Human Organization*, 68, pp. 82–88.

Neria, Y. and Shultz, J.M. (2012) Mental health effects of Hurricane Sandy: Characteristics, potential aftermath, and response. *JAMA*, 308, pp. 2571–2572.

Ng, M., Diaz, R. and Behr, J. (2015) Departure time choice behavior for hurricane evacuation planning: The case of the understudied medically fragile population. *Transportation Research Part E: Logistics and Transportation Review*, 77, pp. 215–226.

Oliver, M. (1990) *The politics of disablement*. London: Palgrave Macmillan.

Oliver, M. (1996) *Understanding disability: From theory to practice*. Basingstoke: Macmillan.

Parnes, P., Cameron, D., Christie, N., Cockburn, L., Hashemi, G. and Yoshida, K. (2009) Disability in low-income countries: Issues and implications. *Disability and Rehabilitation*, 31, pp. 1170–1180.

Pasciak, A. and Kelley, T. (2013) Conformity to traditional gender norms by male police officers exposed to trauma: Implications for critical incident stress debriefing. *Applied Psychology in Criminal Justice*, 9, pp. 137–156.

Phibbs, S., Good, G., Severinsen, C., Woodbury, E. and Williamson, K. (2015) Emergency preparedness and perceptions of vulnerability among disabled people following the Christchurch earthquakes: Applying lessons learnt to the Hyogo Framework For Action. *Australasian Journal of Disaster and Trauma Studies*, 19, pp. 37–46.

Rahman, M.S. and Mallick, M.S. (2007) Community, disability and response to disaster mitigation in Bangladesh. Available from http://www.Publicsphereproject.Org/Events/ Diac08/Proceedings/16.Disaster_Mitigation.Rahman_And_Mallick.Pdf (accessed September 22, 2015).

Reeve, D. (2008) Towards a psychology of disability: The emotional effects of living in a disabling society. In: Goodley, D. and Lawthom, R. (eds.) *Disability and psychology: Critical introductions and reflections*. London: Palgrave, pp. 94–107.

Rosenfield, D. and Faircloth, C.A. (2006) *Medicalized masculinities*. Philadelphia: Temple University Press.

Sherry, M. (2010) *Disability hate crimes: Does anyone really hate disabled people?* Farnham, Surrey, England: Ashgate.

Shuttleworth, R. (2004) Disabled masculinity: Expanding the masculine repertoire. In: Smith, B.G. and Hutchison, B. (eds.) *Gendering disability*. New Brunswick, NJ: Rutgers University Press, pp. 166–180.

Siebers, T. (2010) *Disability aesthetics*. Ann Arbor: University of Michigan Press.

Singh, P. (2007) Impact of the South Asian earthquake on disabled people in the state of Jammu and Kashmir. *The Review of Disability Studies*. 3(3). Available from http://www.rds.hawaii.edu/ojs/index.php/journal/article/view/288/894 (accessed December 22, 2015).

Snyder, S.L. and Mitchell, D.T. (2010) Introduction: Ablenationalism and the geo-politics of disability. *Journal of Literary & Cultural Disability Studies*, 4, pp. 113–125.

Stevens, G.J., Dunsmore, J.C., Agho, K.E., Taylor, M.R., Jones, A.L., van Ritten, J.J. and Raphael, B. (2013) Long-term health and wellbeing of people affected by the 2002 Bali bombing. *Medical Journal of Australia*, 198(5), pp. 273–277.

Takahashi, A., Watanabe, K., Oshima, M., Shimada, H. and Ozawa, A. (1997) The effect of the disaster caused by the Great Hanshin Earthquake on people with intellectual disability. *Journal of Intellectual Disability Research*, 41, pp. 193–196.

Torrell, M.R. (2013) Potentialities: Toward a transformative theory of disabled masculinities. In: Wappett, M. and Arndt, K. (eds.) *Emerging perspectives on disability studies*. New York: Palgrave Macmillan.

Tyler, M. and Fairbrother, P. (2013) Bushfires are "men's business": The importance of gender and rural hegemonic masculinity. *Journal of Rural Studies*, 30, pp. 110–119.

Vaughn, J. (2003) *Disabled rights: American disability policy and the fight for equality*. Washington, DC: Georgetown University Press.

Wilson, N.J., Shuttleworth, R., Stancliffe, R. and Parmenter, T. (2012) Masculinity theory in applied research with men and boys with intellectual disability. *Intellectual and Developmental Disabilities*, 50, pp. 261–272.

Zabeida, N. (2010) Not making excuses: Functions of rape as a tool in ethno-nationalist wars. In: Chandler, R.M., Wang, L. and Fuller, L.K. (eds.) *Women, war and violence: Personal perspectives and global activism*. New York: Palgrave Macmillan, pp. 17–30.

11 Masculinity, sexuality and disaster

Unpacking gendered LGBT experiences in the 2011 Brisbane floods in Queensland, Australia

Andrew Gorman-Murray, Scott McKinnon and Dale Dominey-Howes

Lesbian, gay, bisexual and transgender (LGBT) populations experience forms of vulnerability and resilience in disaster contexts tied specifically to their sexual and gender minority status. Discriminatory government policies and emergency management practices, for example, may leave LGBT individuals, couples and families more vulnerable to disaster impacts (Balgos *et al.*, 2012; Cianfarani, 2013; Dominey-Howes *et al.*, 2014). In this chapter, we investigate intersections between LGBT experiences of disaster and masculinities. Our case study focuses on some experiences of LGBT populations in Brisbane, Australia, during the floods of January 2011. Drawing on data from survey responses, semistructured interviews and media analysis, we suggest that the vulnerability of lesbian and transgender populations may be exacerbated by the privileging of gay male voices within disaster responses by LGBT communities (Browne, 2007; Browne *et al.*, 2010). We also argue, however, that certain performances and expectations of masculinity within the disaster context may marginalize gay men (Coston and Kimmel, 2012). Our concern is with the complex ways in which masculinities interact with, and are produced within, the situation of a disaster (Pease, 2014).

LGBT populations in Australia are a marginalized group who experience discrimination and peripheralization in everyday life (Leonard *et al.*, 2012). It is important to note, however, that marginality is not experienced evenly across the disparate lives encompassed by the LGBT acronym. Moreover, intersections of age, dis/ability, race and class also modulate marginality and vulnerability, including within disaster contexts (D'Ooge, 2008; Pincha and Krishna, 2008). Our interest in this chapter lies in understanding how masculinities may operate in and through LGBT experiences of disaster. Rather than arguing that hegemonic masculinity results in uniform forms of vulnerability for LGBT populations, we suggest that a more nuanced understanding reveals both differing impacts of vulnerability, as well as ways in which hegemonic masculinity may in fact be resisted or deployed as a means of resilience.

In the following section, we describe our conceptual framework. We then provide an overview of data and methods and briefly describe the Brisbane floods of January 2011. The substantive discussion of the chapter unfolds in four sections. First, we describe the effects of LGBT marginalization during the disaster. Second, we discuss how gay men experienced their masculinity and encountered hegemonic masculinity during the disaster. Third, we examine how both gay men and lesbians negotiated hegemonic masculinity in their interactions with male Armed Services personnel deployed in disaster response and recovery. Fourth, we consider the implications of gay male privilege within the LGBT community during the disaster. Finally, we offer a brief conclusion based on our arguments regarding intersections between LGBT identities and masculinities in disaster contexts.

Relational masculinities in place

We understand masculinities as relational and constituted in and through temporal and spatial contexts (Berg and Longhurst, 2003; Hopkins and Noble, 2009). In this section, we provide a brief outline of this conceptual framework and how it relates to our arguments in this chapter.

Although masculinity and femininity were once understood as fixed binary categories, the work of feminist scholars in the 1960s and 1970s began to challenge these strict definitions, making clear the social dimensions of how gender is understood, experienced and performed (Gorman-Murray and Hopkins, 2014). The work of sociologists in the field of critical men's studies (Connell, 2005) and geographers exploring geographies of masculinities (Jackson, 1991) subsequently began to define masculinity as a range of shifting modes of being (hence the plural "masculinities"). Masculinity, both as an ideology and identity, is thus understood as constructed in relation to categories including gender, age, race, sexuality and dis/ability (Berg and Longhurst, 2003; Hopkins and Noble, 2009). Moreover, these interactions are contingent upon the time and space in which they occur.

Within a gendered hierarchy of power relationships, the concept of "hegemonic masculinity" has been posited as a normative archetype, a form of masculine ideal toward which men may aspire but which is rarely achievable (Connell, 2005). Although this concept has received valuable critique (Moller, 2007), it nonetheless provides a useful framework for thinking about gender as both hierarchical and performative. As Gorman-Murray and Hopkins argue, "Maintaining the gender order and its power through the hegemonic masculine ideal relies not only on the subordination of all women and feminine subjectivities, but also the subordination of most masculine subjectivities to this ideal" (2014, p. 7). For example—and pertinent to our analysis—heteronormativity and homophobia are central to the hegemonic ideal, and nonheterosexual men (gay, bisexual, queer) are among the most marginal subjects within hierarchies of masculinities (Connell, 1992, 2005).

In this chapter, we examine how masculinities have been constructed, resisted and produced in spaces impacted by a disaster by contemplating interactions with sexual and gender minority identities. Significantly, our aim is not to provide an uncritical analysis of hegemonic masculinity as uniformly producing marginality among minority identities, but rather to suggest the multiplicity of possibilities in interactions between identities constituted in space and time (Enarson, 2010). Although we argue that the dominance of masculine voices has had specific impacts on how lesbians, bisexual women and transwomen experience disaster, as well as on interactions between gay and heterosexual men, we also find nuance and complexity in how marginalized identity groups may resist masculine privilege and ultimately exhibit resilience.

Project, data and methods

The data were drawn from two sources: first, an ongoing research project being undertaken by the authors into the impacts of disasters on LGBT populations in Australia and New Zealand; and second, a survey conducted by a LGBT community health organization, Queensland Association for Healthy Communities (QAHC), in response to the 2011 Queensland floods. The QAHC survey comprised 70 open and closed questions. With a total of 48 completed responses, the survey is not representative but offers insights into the experiences of the LGBT community. Of those respondents, 44% identified as female, 33% male and 23% transgender or genderqueer. The survey recorded impacts across Brisbane and South East Queensland, with 79% of respondents living in Brisbane.

Our ongoing project takes a mixed-methods approach including media analysis, online surveys and semistructured interviews. Eight LGBT individuals in Brisbane have been interviewed and in this chapter we draw on in-depth analysis of three interviews. We believe this approach offers an opportunity not only to draw out commonalities of experience, when considered in the context of the QAHC survey results, but also to contemplate the specificity of individual experiences within disaster contexts. As such, we provide critical analysis of interviews with one gay man, one lesbian and one transwoman.

We also draw on content analysis of media reporting of the Brisbane floods. In that analysis, we investigated the reporting of the floods in Australian media publications created by and for the LGBT community and which were accessible to Brisbane's LGBT population. Our interest was in examining, firstly, the interest taken by the LGBT media in reporting on disaster experiences specific to the LGBT community and, secondly, the equity of gender representation in reporting of the floods. We searched the publications *QNews, Gay News Network, Star Observer* and *Lesbians on the Loose* (*LOTL*) for any reporting of the Brisbane floods from January 2011 until September 2012. We then conducted content analysis of the articles ($n = 21$) to obtain quantitative data on the gender and sexual identity of informants included in flood reports.

2011 Brisbane floods

Extremely heavy rainfall across December 2010 and January 2011 in Queensland resulted in widespread flooding. In Brisbane, the capital city, the city's flood gauge exceeded its major flood level on January 12 (Queensland Floods Commission, 2011, p. 27). At the flood's peak on January 13, 14,100 properties were affected and 1,203 houses inundated. Although no deaths were recorded in Brisbane itself, flooding from January 10 to 24 resulted in 22 deaths in South East Queensland (Martin, 2011). Thirty-four emergency shelters were established to accommodate approximately 12,000 displaced residents. Calls for volunteer assistance for flood victims were met with an extraordinary response. On the weekend of January 15–16, between 50,000 and 60,000 volunteers, denoted "the Mud Army," are believed to have participated, helping to remove mud, clear away debris and discard destroyed belongings (Rafter, 2013).

Queering the Brisbane floods: LGBT marginality and vulnerability

Social marginality is a key factor determining vulnerability and resilience in disaster contexts (Wisner, 2008; Reid, 2013). Research at a number of sites globally has revealed ways in which LGBT identity or status has contributed to the vulnerability of LGBT populations impacted by disaster (D'Ooge, 2008; Gaillard, 2011; Dominey-Howes *et al.*, 2014; Gorman-Murray *et al.*, 2014). While we are unable to provide a detailed overview of that research here, we briefly discuss one exemplary consequence of marginality for LGBT populations, namely exclusion from necessary support services.

In Australia, LGBT populations continue to experience social marginality (Leonard *et al.*, 2012). LGBT Australians report poorer mental and general health outcomes than the wider population, as well as greater vulnerability to violence and reduced access to services. Disasters generate a need to access a range of services or to interact with a range of "official" institutions, including emergency services personnel, charity groups and government departments. The QAHC survey asked about respondents' comfort or confidence in engaging with these services, seeking to understand whether LGBT status inhibited approaching or accessing necessary support. Over 50% of respondents stated that their anxiety or uncertainty was such that they chose not to access mainstream emergency services (i.e. temporary shelters, referral services, information centers and government agencies). Apprehension about potential discrimination as LGBT individuals, couples or families therefore played a role in increasing vulnerability (Gorman-Murray *et al.*, forthcoming).

Some participants used the open response option to explain further. One stated, "Discrimination when accessing mainstream services is always an issue—you never know if you will be treated properly and with respect." Another wrote, "I would have been concerned my relationship may not have been accepted in mainstream support services." These responses affirm research in other locations overseas indicating uncertainty in interactions with support services, and experiences of discrimination by and exclusion from support agencies (Balgos *et al.*, 2012; Cianfarani, 2013; D'Ooge, 2008; Pincha and Krishna, 2008). Social marginality is thus central to LGBT disaster experiences. Below, we begin to break down the role of gender as, and in relation to, masculinity, and as it both produces further vulnerabilities and is a point of resistance.

Relations between men and across masculinities: gay men and disasters

As men, gay men arguably have access to some forms of masculine privilege (Kendall and Martino, 2006). Yet simultaneously, gay men are marginal in hierarchies of masculinities, and experiences and expressions of heteronormativity, heterosexism and homophobia are critical in determining gay male vulnerability and capacity in disaster events. As Connell argues: "Patriarchal culture has a simple interpretation of gay men: they lack masculinity" (2005, p. 143). In many ways, hegemonic masculinity is defined by its difference from, or exclusion of, male homosexuality. Gay men's gender may, at times, operate as a point of privilege; however, the embodiment and identification of gay masculinity may exacerbate vulnerabilities in disaster contexts. We suggest that encounters between differing masculinities can affect how gay men experience disaster.

Gay men's mental and physical health outcomes are statistically poorer than the general population (Lyons *et al.*, 2014). In interviews, gay men reported ways in which the management of existing health issues was critical to how they experienced the Brisbane floods. Mike and his partner David (60-something Anglo-Australians), for example, live in an upper-floor apartment in the riverside suburb of West End. During the floods, they were trapped in their apartment without power for several days. Mike detailed how David's and his experiences of the disaster were determined by existing health issues. David suffers from a range of physical and mental health problems and, as a result, is unable to work. According to Mike,

> [David] felt like he was a burden on the house because he wasn't contributing anything financially anymore. So that led to the depression becoming more severe. And then the floods came and that was, sort of, not a tipping point but it just added another layer of stress and anxiety to his loss that he didn't really need. I mean I think I'm a pretty resilient and tough person but, um, he's not.

As Nardi argues: "Romantic relationships are another site in which gay men must deal with issues of masculinity" (2000, p. 8). Interactions between men in same-sex relationships demonstrate varied relational masculinities: gay masculinities are multiple. This experience of the flood reveals how masculinities come in to play *between* gay men in the everyday space of the home, and how the added trauma of the disaster may exacerbate preexisting difficulties. David's inability to provide within the home, frequently defined as a critical element of masculine performance (McDowell, 2003), had contributed to his depression, which in turn left him more vulnerable to the traumatic impacts of the disaster.

The contexts of the disaster also produced interactions between gay and heterosexual men, offering insights into the negotiation of hegemonic masculinity by gay men (Gorman-Murray, 2013). In their everyday lives, gay men must negotiate the risk of being "out," potentially deciding to participate in a politics of identity by being out in specific spaces or contexts, or choosing to remain closeted in spaces or circumstances in which the risk appears too high (Linnemann, 2000). The circumstances of a disaster may remove any degree of choice from this process of risk assessment.

Critically, the removal of this choice may also occur within the space of the home. Domestic spaces are important locations of ontological security and well-being for gay men, in which some degree of control can be maintained over who is allowed in or not (Gorman-Murray, 2012). Again, a disaster can remove that choice, with emergency workers, support services and volunteers all coming into the residence. Control over the boundaries of the home, and thus security of self, is compromised. Mike explained the experience:

> Uh, it was really strange having this bunch of people I've never seen in my life rifling through all my personal belongings. And whilst I'm not closeted in any way, shape or form, it still felt strange to have people digging out old photo albums and going through them and saying, "Do you want to keep this? Oh, look at those photos."

It is interesting to note that Mike specifically acknowledges the role of sexuality. Although he is "not closeted," he nonetheless describes a sense in which the removal of choice about revealing (or not) his sexuality in this spatial and temporal circumstance is experienced as a form of vulnerability. Many of the photographs, books and other documents damaged by the floodwaters, and which volunteers were helping to clean or throw away, were likely to identify Mike and David as a gay couple. This led to a degree of apprehension related to the intersection of gender and sexuality in and beyond the (damaged) home. Here, gay masculinities are understood as constituted and supported within the domestic space of the same-sex family household, but compromised and potentially threatened by the perceived heterosexual masculinity of those invading that space (albeit with good intentions).

Gay men and lesbians encountering and resisting hegemonic masculinity in disasters

In the Brisbane floods, as in many disasters, Armed Services personnel were deployed to assist with response and recovery, including help with securing, repairing and cleaning out affected homes. Since the Armed Services are "a significant reservoir for the articulation of masculinity within society at large" (Atherton, 2014, p. 143), the interactions of gay men and other LGBT people with male personnel offer insight into the operation of masculinities within disasters, including how LGBT individuals encounter and contest hegemonic masculinity.

Mike's interactions with male soldiers reveal how hegemonic masculinity can be resisted in ways that are ultimately empowering. As Connell argues:

> Heterosexual masculinity … is encountered in the form of everyday relations with straight men that often have an undercurrent of threat … But this does

not mean conceding legitimacy. Straight men may also be seen as the pathetic bearers of outmoded ideas and a boring way of life.

(2005, p. 155)

Although "pathetic" and "boring" are overly strong terms to describe the story told by Mike, there was certainly playful humor found in the ignorance and naivety of some young, male soldiers. When these soldiers were helping to clean Mike's garage:

> the first [thing] they found was this old canvas tote bag which was where my partner and I stored our out-of-date VHS porn tapes. And, um, given that our taste in those days was to a military bent, these boys found all these VHS tapes and none of them clicked that they were porn. All they clicked on to was the fact that they were men in military uniforms. And the military uniforms weren't quite right, you know, the designations or the colours weren't right, or the titles, or you know. It was all I could do [not to] burst into hysterical laughter.

The soldiers' ignorance of the content of the videos, and their concern for the incorrect insignia on the uniforms of the men on the video covers, were deployed as a moment of humor that countered the apprehension involved in having heterosexual male strangers invade the home. A form of gendered knowledge and identity work was enacted to alleviate Mike's tension about potential homophobia, or at least uneasy relations between heterosexual and homosexual men, in stressful and traumatic post-disaster circumstances.

Similarly, Anne's interactions with male soldiers demonstrate how lesbians also negotiated masculine privilege in the disaster, and give further insight into the instabilities and paradoxes of hegemonic masculinity in this context. In her 50s and living alone in an outer suburb, her home was inundated by floodwaters, which reached almost to the ceiling of the second story. While she had formerly worked as a carpenter, and had the tools and skills for doing repair and recovery tasks, as an older, single person the disaster overwhelmed her ability to cope. She sought help from the Armed Services personnel sent to assist. This, however, is not a story of a woman being rescued by strong men.

In the days after the flood, Anne noticed a group of young, male soldiers sitting in a truck parked near her home, and twice requested their assistance with clearing away mud. They replied that they could not help because, as one explained, "They would get mud on their boots." Displeased with this response, Anne's next step is worth quoting at length:

> So when their truck came past me on the street, I went and stood out in the middle of the street, in my muddy gumboots and dirty shorts, put my hands on my hips and stared at them, right in the face. … And so they came to a halt beside me and got out of their truck and said, "Alright, what is it you would like?" And I said, "This is the mud, I've explained before. Could you please

grab these tools, which I can lend you … and all I would like for you to do please, because it's just too much for me with my sore back, is just to put it in the backyard." So they gave the command, and they all came out of the back of the truck and they did it.

There is a sense in which her self-depiction as dirty, sore and struggling as she cleans away the mud, while men watch with their clean boots from the safety of their truck, imbricates vulnerability and resilience. Gender becomes complex in this tale, with the performance of masculinity made to look ridiculous under Anne's satirical gaze: this is a story of a strong woman demanding work from men whose masculinity is arguably compromised by their concern for the state of their boots! As with Mike's experience, hegemonic masculinity is resisted and undermined here in a way that ultimately adds to personal resilience.

Negotiating gay male privilege within the LGBT community in disasters

Even while gay men, lesbians and other sexual and gender minorities might encounter and contest hegemonic masculinity and gender hierarchies in the context of the disaster, gay masculinity can still emerge as a privileged identification *within* LGBT disaster experiences. Our media analysis suggests this may have happened in the Brisbane floods. This is problematic as it recenters men's experiences and masculinist action and knowledge even within a marginalized LGBT community, and may deflect or diminish recognition of the specific gendered needs of lesbians, bisexual women and transwomen in the disaster (Enarson, 2010; Pease, 2014). Consequently, accounting for the full spectrum of LGBT disaster experiences and needs means remaining sensitive to, and actively questioning, gay male privilege within LGBT communities.

From the date of the floods until September 2012, a total of 21 articles appeared in the LGBT media about the Brisbane floods, with the majority appearing within two months of the event. Analysis of gender inclusivity suggests unequal representation of LGBT disaster experiences in that reporting. Of the 21 articles, only four (19%) included a female informant. This finding accords with earlier research noting the dominant voice of gay men in the Queensland LGBT media (Robinson, 2007). This suggests the possibility that within experiences of LGBT marginality, lesbians, bisexual women and transwomen are further marginalized by their gender. The overwhelming dominance of gay male voices in the LGBT media occluded a range of experiences particular to women, as well as to queer individuals whose gender identity does not fit within a male/female binary, which were largely overlooked.

Indeed, in our interviews, some women, especially lesbians and transwomen, described experiences during the floods that were specific to their gender identities, experiences absent from broader disaster narratives dominated by male voices. Carolyn is a twenty-something, Anglo-Australian, transgender woman who was living in a share-house in inner-city West End when the flood struck.

The flood inundated Carolyn's home and she was forced to find alternative accommodation with friends and work colleagues for several weeks. Losing the safe space of the home entailed particular gender-related challenges:

> for me, being a trans person … needing a sense of safety, privacy. Um, 'cause it was quite recent, I—maybe the year before I'd just had surgery, ah lower surgery, so I was still really recovering from that. … [It] felt really tricky … finding a safe space to recover.

The loss of the safe space of the home is a common disaster-related impact that must be negotiated regardless of sexuality or gender. Added vulnerabilities tied to minority sexual or gender identity, however, may exacerbate this loss (Gorman-Murray *et al.*, 2014). Carolyn makes clear some of the specific difficulties that may arise. The surgery she discusses is related to the process of gender reassignment. That difficult process of recovery could be most successfully negotiated within the safety of her home. When the disaster removed access to that space, her process of recovery became more challenging. Carolyn's disaster narrative thus reveals vulnerabilities not experienced by gay men and potentially left invisible by the dominance of gay male voices in how LGBT disaster experiences are understood.

Although the dominance of gay male voices in LGBT reporting may contribute to the vulnerability of lesbians, bisexual women and transwomen, it is equally important to note that friendship and community networks also operated *across* gender difference within LGBT communities, that is, between lesbians, transwomen, bisexual people and gay men. These LGBT solidarities facilitated the resilience and coping capacities of flood victims. Carolyn stayed with a gay male friend for several weeks while her home was repaired. Belinda's gay male friends stored furniture for her to avoid flood damage. Another interviewee, Brian, is involved in an LGBT-friendly church and the members of that congregation provided enormous support to one another through the disaster. The narratives of LGBT people impacted by the Brisbane floods thus reflect both the challenges of negotiating masculinities and gender difference, as well as the ways in which LGBT communities seek to build resilience and resist marginality via established social networks and social capital (Weeks *et al.*, 2001).

Conclusion

By investigating the experiences of LGBT populations in the 2011 Brisbane floods, we have revealed the complex and contingent operation of relational masculinities in disaster contexts. We have shown that social marginality has specific consequences on the experience of disasters for LGBT individuals, couples and families. Gay men may experience masculinity as a site of both privilege and vulnerability. When negotiating masculine privilege in disaster contexts, lesbians, bisexual women and transwomen may find their experiences left invisible, but may also develop strategies through which to resist that privilege.

Marginality must be understood as unevenly experienced, as producing vulnerability as well as being resisted or negotiated in ways that build capacity. In understanding the operation of masculinity and sexual or gender minority status in disaster contexts, therefore, it is necessary to both acknowledge the critical role of gender in producing vulnerability, while allowing for a nuanced understanding of the relational and spatial constitution of these various identities. Further research may begin to unpack the different performances of masculinities activated and interacting in the disaster context and their effects on response and recovery.

References

Atherton, S. (2014) The geographies of military inculcation and domesticity: Reconceptualising masculinities in the home. In: Gorman-Murray, A. and Hopkins, P. (eds.) *Masculinities and place*. Farnham: Ashgate, pp. 143–157.

Balgos, B., Gaillard, J.C. and Sanz, K. (2012) The *warias* of Indonesia in disaster risk reduction: The case of the 2010 Mt Merapi eruption. *Gender and Development*, 20(2), pp. 337–348.

Berg, L. and Longhurst, R. (2003) Placing masculinities and geography. *Gender, Place and Culture*, 10(4), pp. 351–360.

Browne, K. (2007) Lesbian geographies. *Social and Cultural Geography*, 8(1), pp. 1–7.

Browne, K., Nash, C.J. and Hines, S. (2010) Towards trans geographies. *Gender, Place and Culture*, 17(5), pp. 573–577.

Cianfarani, M. (2013) Supporting the resilience of LGBTQI2S people and households in Canadian disaster and emergency management. MA. Royal Roads University.

Connell, R.W. (1992) A very straight gay: Masculinity, homosexual experience, and the dynamics of gender. *American Sociological Review*, 57(6), pp. 735–751.

Connell, R.W. (2005) *Masculinities*. Crow's Nest, NSW: Allen & Unwin.

Coston, B.M. and Kimmel, M. (2012) Seeing privilege where it isn't: Marginalized masculinities and the intersectionality of privilege. *Journal of Social Issues*, 68, pp. 97–111.

Dominey-Howes, D., Gorman-Murray, A. and McKinnon, S. (2014) Queering disasters: On the need to account for LGBTI experiences in natural disaster contexts. *Gender, Place and Culture*, 21(7), pp. 905–918.

D'Ooge, C. (2008) Queer Katrina: Gender and sexual orientation matters in the aftermath of the disaster. In: Willinger, B. (ed.) *Katrina and the women of New Orleans*. New Orleans: Tulane University, pp. 22–24. Available from http://tulane.edu/nccrow/publications.cfm (accessed November 10, 2015).

Enarson, E. (2010) Gender. In: Phillips, B., Thomas, D., Fothergill, A. and Blinn-Pike, L. (eds.) *Social vulnerability to disasters*. Boca Raton: CRC Press, pp. 123–154.

Gaillard, J.C. (2011) *People's response to disasters: Vulnerability, capacities, resilience in the Philippine Context*. Angeles City, Pampanga: Center for Kapampangan Studies.

Gorman-Murray, A. (2012) Queer politics at home: Gay men's management of the public/private boundary. *New Zealand Geographer*, 68(2), pp. 111–120.

Gorman-Murray, A. (2013) Straight–gay friendships: Relational masculinities and equalities landscapes in Sydney, Australia. *Geoforum*, 49, pp. 214–223.

Gorman-Murray, A. and Hopkins, P. (2014) Introduction: Masculinities and place. In: Gorman-Murray, A. and Hopkins, P. (eds.) *Masculinities and place*. Farnham: Ashgate, pp. 1–24.

Gorman-Murray, A., McKinnon, S. and Dominey-Howes, D. (2014) Queer domicide? LGBT displacement and home loss in natural disaster impact, response and recovery. *Home Cultures*, 11(2), pp. 237–262.

Gorman-Murray, A., Morris, S., Keppel, J., McKinnon S. and Dominey-Howes, D. (forthcoming) Problems and possibilities on the margins: LGBT experiences in the 2011 Queensland floods. *Gender, Place and Culture*.

Hopkins, P. and Noble, G. (2009) Masculinities in place: Situated identities, relations and intersectionality. *Social and Cultural Geography*, 10(8), pp. 811–819.

Jackson, P. (1991) The cultural politics of masculinity: Towards a social geography. *Transactions of the Institute of British Geographers*, 16(2), pp. 199–213.

Kendall, C. and Martino, W. (2006) Introduction. In: Kendall, C. and Martino, W. (eds.) *Gendered outcasts and sexual outlaws: Sexual oppression and gender hierarchies in queer men's lives*. New York: Harrington Park Press, pp. 5–16.

Leonard, W., Pitts, M., Mitchell, A., Lyons, A., Smith, A., Patel, S., Couch, M. and Barrett, A. (2012) *Private lives 2: The second national survey of the health and wellbeing of gay, lesbian, bisexual and transgender (GLBT) Australians*. Melbourne: Australian Research Centre in Sex, Health and Society, La Trobe University.

Linnemann, T.J. (2000) Risk and masculinity in the everyday lives of gay men. In: Nardi, P. (ed.) *Research on men and masculinities series: Gay masculinities*. Thousand Oaks: Sage, pp. 83–100.

Lyons, A., Pitts, M. and Grierson, J. (2014) Sense of coherence as a protective factor for psychological distress among gay men: A prospective cohort study. *Anxiety, Stress and Coping*, 27(6), pp. 662–677.

Martin, L. (2011) Time for Queensland to Heal after the Floods—Queensland Premier Anna Bligh. *Herald-Sun*. January 26. Available from http://www.heraldsun.com.au/news/victoria/time-for-queensland-to-heal-bligh/story-e6frf7l6-1225994612338 (accessed November 10, 2015).

McDowell, L. (2003) *Redundant masculinities? Employment change and white working class youth*. Oxford: Blackwell.

Moller, P. (2007) Exploiting patterns: A critique of hegemonic masculinity. *Journal of Gender Studies*, 16, pp. 263–276.

Nardi, P. (2000) "Anything for a sis, Mary": An introduction to gay masculinities. In: Nardi, P. (ed.) *Research on men and masculinities series: Gay masculinities*. Thousand Oaks: Sage, pp. 1–12.

Pease, B. (2014) Hegemonic masculinity and the gendering of men in disaster management: Implications for social work education. *Advances in Social Work and Welfare Education*, 16(2), pp. 60–72.

Pincha, C. and Krishna, H. (2008) Aravanis: Voiceless victims of the tsunami. *Humanitarian Exchange Magazine*, 41. Available from http://odihpn.org/magazine/aravanis-voiceless-victims-of-the-tsunami/ (accessed November 10, 2015).

QNews (2011) QNews flood relief and billet program. Available from http://www.qnews.com.au/article/qnews-flood-relief-billet-program/ (accessed November 10, 2015).

Queensland Floods Commission of Inquiry (2011) *Interim report*. Brisbane. Available from http://www.floodcommission.qld.gov.au/publications/interim-report. (accessed November 10, 2015).

Rafter, F. (2013) Volunteers as agents of co-production: "Mud armies" in emergency services. In: Lindquist, E.A, Vincent, S. and Wanna, J. (eds.) *Putting citizens first: Engagement in policy and service delivery for the 21st century*. Canberra: ANU E-Press.

Reid, M. (2013) Disasters and social inequalities. *Sociology Compass*, 7(11), pp. 984–997.

Robinson, S. (2007) Queensland's queer press. *Queensland Review*, 14(2), pp. 59–78.

Weeks, J., Heaphy, B. and Donovan, C. (2001) *Same-sex intimacies: Families of choice and other life experiments*. London: Routledge.

Wisner, B. (1998) Marginality and vulnerability: Why the homeless of Tokyo don't "count" in disaster preparations. *Applied Geography*, 18(1), pp. 25–33.

12 Indigenous masculinities in a changing climate

Vulnerability and resilience in the United States

Kirsten Vinyeta, Kyle Powys Whyte and Kathy Lynn

Gender shapes Indigenous vulnerability and resilience due to the coupled social and ecological challenges of climate change in Indigenous communities in the United States (Maynard, 1998; Grossman and Parker, 2012; Bennett *et al.*, 2014; Maldonado *et al.*, 2014; Whyte, 2014). Despite its relevance, little research has analyzed the ways in which gender shapes climate change experiences. Even less research has focused on the impacts of climate change on Indigenous masculinity. With this backdrop, we foreground Indigenous men and masculinities with respect to climate change vulnerability and resilience.[1]

We open this chapter by briefly describing pre-contact Indigenous conceptions of gender in the US, followed by a discussion of how settlement has affected gender roles, relations and gendered traditional knowledge in Indigenous communities. We then describe some of the ways in which Indigeneity and masculinity are intersecting (or may intersect) with climate change in four key arenas: health, migration and displacement, economic and professional development, and culture. We follow this with a discussion of Indigenous men's roles in political resistance and climate change resilience. We conclude by summarizing the key implications for Indigenous climate change initiatives and for the ongoing reconstruction and reassertion of Indigenous gender identities. This discussion draws on peer-reviewed and grey literatures in English focusing on the United States and Canada, and builds on a 2015 literature synthesis by the authors in which we explored how gendered Indigeneity may influence climate change vulnerability and resilience in Indigenous communities in the United States (Vinyeta *et al.*, 2015).

Colonialism and gender systems in Indigenous communities in the United States

Gender has historically played an important role in defining social structure and sociocultural responsibilities within Indigenous communities in North America (Kuhlmann, 1992; Jacobs *et al.*,1997; Roscoe, 1998; Anderson, 2005; McGregor, 2005). Traditionally, responsibilities to land, water, plants and animals were gendered. In some Indigenous communities, women were responsible for managing and harvesting plants and engaging in agricultural activities, whereas men were responsible for hunting and/or fishing activities (Kuhlmann,1992; Anderson,

2005; Scarry and Scarry, 2005; Colombi, 2012). This gender system positioned men and women differently as stewards of key environmental resources with gendered knowledge as well as gendered connections to the landscape.

Prior to the European colonization of North America, many Indigenous communities were characterized by egalitarian relationships between men and women, as well as by women's leadership (Allen, 1992; Roscoe, 1998; Wagner, 2001; Kauanui, 2008; Weaver, 2009; Brave Heart *et al.*, 2012). The institutionalized presence of more than two genders (Roscoe, 1998; LaFortune, 2010; Morgensen, 2011) and minimal gender violence (Deer, 2009; Weaver, 2009; Brave Heart *et al.*, 2012) further distinguished gender systems before colonization, as did gender diversity.

The gendered responsibilities of gender-variant individuals (known today as Lesbian, Gay, Bisexual, Transgender, Two-Spirit or Queer [LGBTTQ] persons) varied by community. Tribes had their own linguistic terminology for gender that often denoted specific names and responsibilities. In some cases, gender-variant individuals preferred and were respected for excelling at traditional activities and responsibilities of a sex not their own (Roscoe, 1998). Yet gender-variant individuals did not necessarily abandon the traditional responsibilities ascribed to their biological sex; sometimes, they pursued both traditionally masculine and feminine activities and responsibilities (Farrer, 1997; Roscoe, 1998).

In the aftermath of a complex colonial history, and in its midst, Indigenous peoples strive to carry out, as well as reconstruct, traditional gender roles and responsibilities as part of their resurgence as self-determining cultures and nations (Allen, 1992; Calhoun *et al.*, 2007; Green, 2007; Mayer, 2007; Kauanui, 2008; Tengan, 2008; Goeman and Denetdale, 2009; LaFortune, 2010; Rifkin, 2011). Indigenous leadership on issues ranging from sexual violence to climate change is often driven by gender-based activism and gender-based responsibilities (McGregor, 2005; Barker, 2008; Bruce and Harries, 2010; Kenny and Fraser, 2012; Whyte, 2014). As such, climate change initiatives are an important context for Indigenous reconstructions of meaningful and healthy gender roles and relations.

Indigenous men and masculinities in a changing climate

While far from monolithic, the norms and practices of some Indigenous masculinities may make men and boys particularly vulnerable to negative climate change impacts. These vulnerabilities are most likely when the responsibilities and roles of traditional Indigenous masculinities are already compromised by colonization, and when men are further compromised by the ways in which communities adapt to climate change. In a study from the University of Oregon on the impacts of environmental change on Karuk masculinity, Norgaard and her colleagues (forthcoming) write:

> While changing environmental conditions affect the gender performances of women and men in unique ways, they appear to be particularly damaging for men. ... Male identified activities of hunting and fishing have been more acutely impacted by environmental decline and state regulation.

While in some cases, traditional masculinities are associated with male dominance and privilege, Indigenous men must strive to disentangle their masculine identities from Western forms of patriarchy that oppress Indigenous women and LGBTTQ persons. Others suggest that claiming traditional powers through exercising traditional masculine roles and responsibilities may represent a form of resurgence which, in Indigenous communities, is critical both to political sovereignty and to cultural and ecological preservation (Norgaard *et al.*, forthcoming; Tengan, 2008).

These tensions bear directly on the gendered approaches to climate change that are profoundly affecting Indigenous communities in North America and beyond. As these issues are underexplored in current literature, we discuss below the climate change vulnerabilities and opportunities for resilience experienced by Indigenous men, and call for further study to better understand the issues facing Indigenous peoples of all genders.

What is at stake?

Many Indigenous livelihoods are dependent on the interrelationships between environment, economy and culture. Donatuto *et al.* (2014) describe six Indigenous community health indicators to examine community health and well-being in the context of climate change. These indicators, which provide a framework for understanding Indigenous livelihoods, are community connection, natural resource security, cultural use, education, self-determination and balance. We turn now to the impacts of climate change on health, economy and culture in order to elaborate on the many ways that livelihoods among indigenous men are impacted by climate change.

Health

Within American Indian communities, research suggests that men's health is particularly vulnerable. Brave Heart *et al.* (2012) report that death rates of American Indian males exceed those of females for every age group up to age 75, primarily due to such health disparities as cardiovascular disease, cancer and diabetes, as well as to disproportionately high rates of suicide, substance abuse and mental health disorders. In the broader literature on gender and disaster, a 2010 study (Krug *et al.*, 1998) found that the suicide rate increased 21.8% for men as compared with 14.5% for women during the types of disasters and years for which the study researchers found significant post-disaster increases in suicide rates. Climate change may exacerbate these vulnerabilities if it contributes to the driving forces behind these health disparities.

Indigenous men may face risks in the event of severe weather or natural disasters, particularly if they associate their masculinity with heroic acts and risk-taking. Rhoades (2003) has described the prevalence of risk-taking behaviors among American Indian men, and points out that American Indian men are more likely to die from accidents than American Indian women. Explanations for this include loss of cultural identity, loss of traditional male roles, failure of primary socialization and unresolved grief from historical trauma (Evans-Campbell, 2008). Indigenous men are also less likely to utilize outpatient and inpatient services than Indigenous women (Rhoades, 2003), paralleling international trends indicating that men and boys are less likely to seek help for mental health issues (World Health Organization, 2011).

Men are also vulnerable to accidents if the environments in which they carry out their traditional activities (including hunting, whaling and fishing) become more hazardous as a result of climate change. This trend is manifest in the Arctic, where Alaska Natives and other Indigenous groups face unusual risks due to the unpredictable nature and thinning of sea ice on which they traditionally hunt (Ford and Smit, 2004; McBeath and Shepro, 2007). Climate change also threatens to raise water temperatures, thereby negatively affecting cold-water fish such as salmon and trout. Tribes in the Pacific Northwest have already had to cope with reduced salmon availability as a result of hydroelectric dams, pollution and overharvesting (Norgaard, 2004; Hanna, 2007; Johnsen, 2009). Climate change may further challenge tribal access to these critical species, further compromising men's traditional roles, knowledge and activities.

For Indigenous men whose health disparities relative to other populations are notable, access to traditional foods that promote health is particularly important. Climate change may affect the availability and quality of traditional food sources, requiring adaptation strategies that will help Indigenous communities retain access to critical plant and animal species (Lynn *et al.*, 2013). The stewardship, harvest, preparation and consumption of traditional foods involve gendered cultural responsibility and traditional knowledge, calling for consideration of gender in food sovereignty movements (Kuhlmann, 1992; Anderson, 2005; McGregor, 2005; Scarry and Scarry, 2005; Marshall, 2006; Bruce and Harries, 2010; Colombi, 2012).

International research indicates that men's mental health may be particularly vulnerable to environmental changes that affect the places and resources critical to masculine identities, especially those closely tied to livelihoods (Kukarenko, 2011; Norgaard *et al.*, forthcoming; WHO, 2011). The stress, loss and cultural changes associated with colonization, combined with the introduction of alcohol, have led to unusually high rates of substance abuse, suicide and violence within Indigenous communities (Maracle, 1996; Ross, 1998; Mokuau, 2002; Strickland *et al.*, 2006; Weaver, 2009). In her analysis of gendered climate change impacts on human health in the Arctic, Kukarenko (2011, "Arctic research," para 1) notes:

> The disruption of traditional roles for men has been identified in a number of studies as a reason for profound problems in male identity and loss of men's self-esteem, which, in turn, leads to a lot of psycho-social disorders among men, including higher suicide rates and alcoholism.

Indigenous men's mental health vulnerabilities can be particularly detrimental to Indigenous communities if they exacerbate, or fail to address, gender-based violence and oppression that harms Indigenous women and LGBTTQ persons. Indigenous masculinities inevitably interact with, and are shaped by, Western forms of patriarchy and patriarchal oppression that disempower those with feminine identities. Indigenous women experience some of the highest rates of violence of any women in the US, most often at the hands of non-Indigenous men (Amnesty International USA, 2007; Weaver, 2009), making them particularly vulnerable to the escalating rates of domestic and sexual violence that have been documented after natural disasters or weather-related crises (Nellemann *et al.*, 2011; WHO, 2011; Alston, 2013). In addition, Sweet describes how warming of the circumpolar region has increased interest in resource extraction, and the corresponding economic growth has increased Indigenous women's risk of being trafficked (2014). More research is needed assessing the impacts of climate change on Indigenous men and on the risk to boys of being trafficked into sex work or forced labor.

Migration and displacement

Lee (2003) has conducted research in Alaska suggesting that Indigenous men migrating or being displaced from their homelands to more urban settings may find it particularly difficult to maintain their traditional relationships and knowledge. The following description is revealing:

> Women can forage for berries near the roadsides outside Anchorage, or gather clams along the shoreline of the Kenai Peninsula, but when a man moves from the village to the city he forfeits access to, and specialized knowledge of, the hunting and fishing areas he has known all his life.
>
> (Lee, 2003, pp. 586–587)

Chávez (2011) describes the challenges faced by Southern Arizona migrants in the population she terms LGBTQ. LGBTQ patients are challenged in finding healthcare providers who can adequately meet cultural needs while simultaneously meeting unique health needs. Another obstacle included finding adequate housing in a new location that will not discriminate based on race, migrant status, gender or sexual orientation. Chávez noted that when searching for housing, LGBTQ migrants frequently rely on family and friends. This can be difficult when social networks are altered by displacement or migration. Further, homelessness is higher among LGBTQ youth as "somewhere between 20% and 40% of homeless youth are LGBTQ" (Ray, 2006, cited in Chávez, 2011, p. 204). While Indigenous LGBTTQ people's lived experiences may be different from those of LGBTQ migrants and other LGBTQ youth, these trends serve to identify some of the challenges that Indigenous LGBTTQ peoples may face when migrating or being displaced from their communities.

Economic and professional development

While international trends indicate lower rates of education and migration among women, some Indigenous communities in the US are experiencing the opposite.

Indigenous men are affected by colonial trends that both collapsed traditional economies but also barred Indigenous persons from entering new colonial economies. Kleinfeld and Andrews state that "the gender gap favoring females in post-secondary education is both large and increasing among Alaska Natives." This trend has implications for Alaska Native men, who are experiencing more unemployment, more social challenges, lower rates of marriage, and lower rates of political participation (Kleinfeld and Andrews, 2006, p. 433).

The traditional male role among Alaska Natives emphasized skills and virtues for which schooling is irrelevant but which were vital to the community, making the difference between survival and starvation. These traditional skills remain important, partly in providing food from the land but also in providing a sense of cultural continuity and stability. At the same time that the transition to a mixed-wage and subsistence economy is making hunting skills less vital to sheer survival, the communication and quantitative skills that schools provide are becoming more essential to the flourishing of Indigenous communities (Kleinfeld and Andrews, 2006).

Lee (2003) suggests that Alaska Native women are migrating to urban centers with more frequency and for economic reasons. Research is needed to determine how these trends are affecting Alaska Native men socioeconomically, and to gauge whether men's vulnerabilities will be exacerbated as climate change continues to alter the resources and conditions of traditionally masculine roles and livelihoods.

The economic impacts of climate change may be especially difficult for Indigenous LGBTTQ persons for whom gender- and race-based oppressions may intersect, doubly challenging their ability to find employment. As Sangganjanavanich (2009, p. 128) states, "Although there are laws to protect disadvantaged groups such as women, ethnic groups, and sexual minorities against employment discrimination, this discrimination is still present in various employment settings." For transgender individuals, retaining a job or seeking new employment can be a particularly challenging feat (Sangganjanavanich, 2009) and one more likely to be confronted as employment options are altered by the impacts of a changing climate.

Culture

Wildlife is vital to men's traditional cultural responsibilities in many Indigenous communities in the US (Marshall, 2006; Colombi, 2012). For the Nez Perce Tribe, for example, salmon fishing is a defining characteristic in the community at large but especially central to men's traditional cultural roles and relationships (Marshall, 2006). Writing on the importance of salmon fishing for Nez Perce men and boys, Marshall observes:

> Task groups devoted to fishing are composed primarily of males and are important for developing gender identity and demonstrating a man's ability to contribute to the community. Task groups are significant for teaching young men basic Nez Perce values and world views; socializing them into

adult male roles; teaching them many practical arts; and educating them in family, community, and tribal history.

(2006, p. 773)

For Indigenous men who are displaced or forced to migrate away from their tribal lands, exercising traditional cultural practices may be particularly difficult. This in turn may challenge the continued use, adaptation and transmittance of gendered traditional knowledge that comes with continued relationships with culturally vital species (Lee, 2003).

In preparation for climate change impacts, some tribes may choose to encourage tribal members, particularly within the younger generations, to step outside traditional gendered responsibilities and learn skills and knowledge that would have traditionally been the responsibility of other genders (Swinomish, 2010; Jacob, 2013). For example, the Confederated Tribes of the Umatilla Indian Reservation (CTUIR) are working to preserve cultural practices and skills surrounding traditional foods by encouraging men and women to share nontraditional gender responsibilities (Swinomish, 2010, p. 25):

> Historically, the men harvested and presented the salmon and the deer. The women collected and presented the roots and berries. Today, because climate change is already affecting the availability of foods, some Umatilla tribal members teach their sons *and* daughters to collect both foods.

Indigenous LGBTTQ persons who identify as masculine may exercise their gender identities in a variety of ways.

> In many cases the traditions of gender variance have been forgotten or repressed. Most of the data indicate that very few individuals who live in the role of the opposite sex and who other members of their community classify as belonging to an alternative gender still live on reservations.

(Lang, 1997, p. 108)

Yet Indigenous LGBTTQ identities are being progressively reasserted and strengthened, as is illustrated by the rise of organizations such as the Bay Area American Indian Two-Spirits (BAAITS, 2015).

Political resistance and resilience to climate change: highlighting men's roles

The role of Indigenous masculinities is critical to consider in strengthening community resilience. In a study of the impacts of salmon decline in Northern California on Karuk masculinity (Norgaard *et al.*, forthcoming), the authors explain that Karuk fishermen's ability to harvest fish for their families, their elders and the broader community is a "fundamental expression of the continuity of their culture" as well as of masculine identity. Karuk fishermen not only meet their responsibilities to their families, community and to the fish themselves; they are, in effect, resisting colonial ecological violence and asserting Karuk cultural presence in the river basin. Therefore, when the ability to fish is limited, not only is Karuk masculinity compromised, but the assertion of Karuk sovereignty and expression of resistance to colonialism is also compromised.

Tengan (2008) provides insight into how settler colonial forces have diminished masculine identities, in this case among Native Hawaiians, and analyzes the value of and circumstances leading to the growth of Hale Mua, a group formed by Native Hawaiian men for the purpose of reconstructing and reasserting Indigenous masculine identities. According to Tengan, Hale Mua not only serves to strengthen men's connection to the land, to each other, to their masculine identities and to their culture but also contributes to broader cultural revitalization efforts that strengthen Hawaiian nationalism. In this sense, the changes wrought through climate change affect traditional masculinities in ways that increase both vulnerability and resilience.

These instances illustrate Indigenous masculinities as a complex set of gender roles and responsibilities that have been (and continue to be) compromised by colonial forces, and still play a critical role in Indigenous resurgence and cultural sovereignty efforts. By protecting and promoting the exercise of masculine responsibilities while simultaneously dismantling heteropatriarchal values that disempower Indigenous women and LGBTTQ peoples in climate change and other movements, Indigenous communities can foster health and well-being for their communities as a whole. For these reasons, some Indigenous scholars now call on political institutions to express deference to Indigenous women's collective actions in a climate change context.

Conclusion

This chapter illustrates some of the ways in which Indigenous masculinities may be both vulnerable and resilient in the face of climate change. The literature supports several tentative conclusions that may help motivate and structure future research. First, Indigenous masculinities that are interwoven with Western forms of patriarchy may exacerbate the vulnerability of Indigenous communities by contributing to, or at the very least failing to interfere with, gender-based violence and oppression. Second, the continuance and resurgence of valued traditional masculinities must take climate change into consideration as a way to ensure resilience. Third, masculinities associated with LGBTTQ persons and communities are deeply marginalized, which makes these populations particularly vulnerable in climate adaptation policy and actionable science because their voices are not heard. Lastly, climate change presents an opportunity for Indigenous peoples to reimagine Indigenous gender systems in a way that promotes resurgence in the face of Western patriarchy and enhances community-wide Indigenous resilience.

Positive pathways arise from applying a gender lens to the experiences of men and boys responding to the radically new social, cultural and environmental worlds that will shape their futures and that of their communities in Indigenous America. Many of the vulnerabilities discussed here are common to men and boys in other contexts. However, we should not confuse generalizability with sameness. Each Indigenous community faces unique forms of colonialism and has unique systems of roles, relationships and responsibilities, which shape climate change vulnerability and resilience. Indeed, the literature is only now emerging on gender, Indigenous peoples and climate change. As such, more culturally specific gender analysis is both needed and likely. This will add depth to masculinity studies and to positive strategies for climate adaptation in other marginalized communities.

Note

1 The authors would like to thank the USDA Forest Service Pacific Northwest Research Station for their support of this work.

References

Allen, P.G. (1992) *The sacred hoop: Recovering the feminine in American Indian traditions*. Boston: Beacon Press.

Alston, M. (2013) Introducing gender and climate change: Research, policy and action. In: Alston, M. and Whittenbury, K. (eds.) *Research, action and policy addressing the gendered impacts of climate change*. Dordrecht: Springer, pp. 3–14.

Amnesty International USA (2007) *Maze of injustice: The failure to protect Indigenous women from sexual violence in the USA*. Available from http://www.amnestyusa.org/our-work/issues/women-s-rights/violence-against-women/maze-of-injustice (accessed July 7, 2014).

Anderson, M.K. (2005) *Tending the wild: Native American knowledge and the management of California's natural resources*. Berkeley: University of California Press.

Barker, J. (2008) Gender, sovereignty, rights: Native women's activism against social inequality and violence in Canada. *American Quarterly*, 60(2), pp. 259–266.

Bay Area American Indian Two-Spirits (BAAITS) (2015) About. Available from http://www.baaits.org/#!about/c10fk (accessed May 31, 2015).

Bennett, T.M.B., Maynard, N., Cochran, P., Gough, R., Lynn, K., Maldonado, J., Voggesser, G., Wotkyns, S. and Cozzetto, K. (2014) Indigenous peoples, lands, and resources. Climate change impacts in the United States: The third national climate assessment. In: Melillo, J.M., Richmond, T.C. and Yohe, G.W. (eds.) *U.S. Global change research program*, pp. 297–317. Available from doi: 10.7930/J09G5JR1 (accessed November 9, 2015).

Brave Heart, M.Y.H., Elkins, J., Tafoya, G., Bird, D. and Salvador, M. (2012) Wicasa was'aha: Restoring the traditional strength of American Indian boys and men. *American Journal of Public Health*, 102(S2), pp. 177–183.

Bruce, L. and Harries, K. (2010) The Anishinabe Kweag were bound to protect the water for future generations. *Women & Environments International Magazine*, 82/83, p. 6.

Calhoun, A., Goeman, M., and Tsethlikai, M. (2007) Gender equity for American Indians. In: Klein, S.S. (eds.) *Handbook for achieving gender equity through education*. 2nd edn. Mahwah, NJ: Lawrence Erlbaum Associates, pp. 525–552.

Chávez, K.R. (2011) Identifying the needs of LGBTQ immigrants and refugees in Southern Arizona. *Journal of Homosexuality*, 58(2), pp.189–218.

Colombi, B.J. (2012) Salmon and the adaptive capacity of Nimiipuu (Nez Perce) culture to cope with change. *The American Indian Quarterly*, 36(1), pp. 75–97.

Deer, S. (2009) Decolonizing rape law: A Native feminist synthesis of safety and sovereignty. *Wicazo Sa Review*, Fall, pp. 149–167.

Donatuto, J., Grossman, E., Konovsky, J., Grossman, S. and Campbell, L. (2014) Indigenous community health and climate change: Integrating biophysical and social science indicators. *Coastal Management*, 42(4), pp. 355–373.

Evans-Campbell, T. (2008) Historical trauma in American Indian/Native Alaska communities: A multilevel framework for exploring impacts on individuals, families, and communities. *Journal of Interpersonal Violence*, 23(3), pp. 316–338.

Farrer, C.R. (1997) A "berdache" by any other name...is a brother, friend, lover, spouse: Reflections on a Mescalero Apache singer of ceremonies. In: Jacobs, S.E., Thomas, W. and Lang, S. (eds.) *Two-spirit people: Native American gender identity, sexuality, and spirituality*. Urbana: University of Illinois Press.

Ford, J. and Smit, B. (2004) A framework for assessing the vulnerability of communities in the Canadian Arctic to risks associated with climate change. *Arctic*, 57(4), pp. 389–400.

Goeman, M.R. and Denetdale, J.N. (2009) Native feminisms: Legacies, interventions, and Indigenous sovereignties. *Wicazo Sa Review*, 24(2), pp. 9–13.

Green, J.A. (ed.) (2007) *Making space for Indigenous feminism*. Black Point, NS: Fernwood Publishing.

Grossman, Z. and Parker, A. (eds.) (2012) *Asserting Native resilience: Pacific Rim Indigenous nations face the climate crisis*. Corvallis, OR: Oregon State University Press.

Hanna, J. (2007) "*Oncorhynchus*" *spp*: Climate change, Pacific Northwest tribes, and salmon. *Natural Resources & Environment*, 22(2), pp. 13–17.

Jacob, M.M. (2013) *Yakama rising: Indigenous cultural revitalization, activism, and healing*. Tucson: University of Arizona Press.

Jacobs, S.E., Thomas, W. and Lang, S. (1997) Introduction. In: Jacobs, S.E., Thomas, W. and Lang, S. (eds.) *Two-spirit people: Native American gender identity, sexuality, and spirituality*. Urbana: University of Illinois Press, pp. 1–18.

Johnsen, D. (2009) Salmon, science, and reciprocity on the northwest coast. *Ecology & Society*, 14(2), pp. 1–11.

Kauanui, J.K. (2008) Native Hawaiian decolonization and the politics of gender. *American Quarterly*, 60(2), pp. 281–287.

Kenny, C. and Fraser, T.N. (2012) *Living Indigenous leadership: Native narratives on building strong communities*. Vancouver, BC: University of British Columbia Press.

Kleinfeld, J. and Andrews, J.J. (2006) The gender gap in higher education in Alaska. *Arctic*, 59(4), pp. 428–434.

Krug, E.G., Kresnow, M.J., Peddicord, J.P., Dahlberg, L.L., Powell, K.E., Crosby, A.E. and Annest, J.L. (1998) Suicide after natural disasters. *New England Journal of Medicine*, 338(6), pp. 373–378.

Kuhlmann, A. (1992) American Indian women of the Plains and Northern woodlands. *Mid-American Review of Sociology*, 16(1), pp. 1–28.

Kukarenko, N. (2011) Climate change effects on human health in a gender perspective: Some trends in Arctic research. *Global Health Action*, 4, p. 7913.

LaFortune, R. (2010) Two spirit activism: Mending the sacred hoop. *Tikkun*, 25(4), p. 46.

Lang, S. (1997) Various kinds of two-spirit people: Gender variance and homosexuality in Native American communities. In: Jacobs, S.E., Thomas, W. and Lang, S. (eds.) *Two-spirit people: Native American gender identity, sexuality, and spirituality*. Urbana: University of Illinois Press, pp. 100–118.

Lee, M. (2003) "How will I sew my baskets?": Women vendors, market art, and incipient political activism in Anchorage, Alaska. *American Indian Quarterly*, 27(3/4), pp. 583–592.

Lynn, K., Daigle, J., Hoffman, J., Lake, F., Michelle, N., Ranco, D., Viles, C., Voggesser, G. and Williams, P. (2013) The impacts of climate change on tribal traditional foods. *Climatic Change*, 120(3), pp. 545–555.

Maldonado, J.K., Colombi, B. and Pandya, R. (eds.) (2014) *Climate change and Indigenous peoples in the United States: Impacts, experiences and actions*. New York: Springer.

Maracle, L. (1996) *I am woman: A Native perspective on sociology and feminism*. Vancouver, BC: Press Gang Publishers.

Marshall, A.G. (2006) Fish, water, and Nez Perce life. *Idaho Law Review*, 42, pp. 763–793. Available from http://heinonline.org/HOL/Page?handle=hein.journals/idlr42&div=26&g_sent=1&collection=journals (accessed August 21, 2013).

Mayer, L. (2007) A return to reciprocity. *Hypatia*, 22(3), pp. 22–42.

Maynard, N. (1998) Native peoples-Native homelands climate change workshop: Lessons learned. Paper presented at AGU Fall Meeting Abstracts. Available from http://downloads.globalchange.gov/nca/nca1/native.pdf (accessed August 21, 2013).

McBeath, J. and Shepro, C.E. (2007) The effects of environmental change on an Arctic native community: Evaluation using local cultural perceptions. *American Indian Quarterly*, 31(1), pp. 44–65.

McGregor, D. (2005) Traditional ecological knowledge: An Anishnabe woman's perspective. *Atlantis: Critical Studies in Gender, Culture & Social Justice*, 29(2), pp. 103–109.

Mokuau, N. (2002) Culturally based interventions for substance use and child abuse among Native Hawaiians. *Public Health Reports*, 117, pp. S82–S87.

Morgensen, S.L. (2011) *Spaces between us: Queer settler colonialism and Indigenous decolonization*. Minneapolis: University of Minnesota Press.

Nellemann, C., Verma, R. and Hislop, L. (eds.) (2011) Women at the frontline of climate change: Gender risks and hopes. A rapid response assessment. United Nations Environment Programme, GRID-Arendal. Available from http://www.unep.org/pdf/rra_gender_screen.pdf (accessed February 17, 2016).

Norgaard, K.M. (2004) The effects of altered diet on the health of the Karuk people: A preliminary report. Available from http://karuk.us/press/2005/Health Effects of Altered Diet.pdf (accessed August 21, 2013).

Norgaard, K.M., Reed, R. and Bacon, J. (forthcoming) Environmental decline and changing gender practices: What happens to Karuk masculinity when there are no fish?

Rhoades, E. (2003) The health status of American Indian and Alaska Native Males. *American Journal of Public Health*, 93(5), pp. 774–778.

Rifkin, M. (2011) *When did Indians become straight?: Kinship, the history of sexuality, and Native sovereignty*. New York: Oxford University Press.

Roscoe, W. (1998) *Changing ones: Third and fourth genders in Native North America*. New York: St. Martin's Press.

Ross, L. (1998) *Inventing the savage: The social construction of Native American criminality*. Austin: University of Texas Press.

Sangganjanavanich, V.F. (2009) Career development practitioners as advocates for transgender individuals: Understanding gender transition. *Journal of Employment Counseling*, 46(3), pp. 128–135.

Scarry, C.M. and Scarry, J. (2005) Native American "garden agriculture' in Southeastern North America. *World Archaeology*. 37(2), pp. 259–274.

Strickland, C.J., Walsh, E. and Cooper. M. (2006) Healing fractured families: Parents' and elders' perspectives on the impact of colonization and youth suicide prevention in a Pacific Northwest American Indian tribe. *Journal of Transcultural Nursing*, 17(1), pp. 5–12.

Sweet, V. (2014) Rising waters, rising threats: The human trafficking of Indigenous women in the circumpolar region of the United States and Canada. MSU Legal Studies Research Paper (12-01). Available from http://papers.ssrn.com/sol3/papers.cfm?abstract_id=2399074## (accessed December 10, 2015).

Swinomish Indian Tribal Community (Swinomish) (2010) Swinomish climate change initiative climate adaptation action plan. Available from http://www.swinomish-nsn.gov/climate_change/Docs/SITC_CC_AdaptationActionPlan_complete.pdf (accessed August 9, 2013).

Tengan, T.P.K. (2008) *Native men remade: Gender and nation in contemporary Hawai'i.* Durham, NC: Duke University Press.

Vinyeta, K., Whyte, K.P. and Lynn, K. (2015) *Climate change through an intersectional lens: Gendered vulnerability and resilience in Indigenous communities in the United States.* Gen. Tech. Rep. PNW-GTR-XXX. Portland, OR: U.S. Department of Agriculture, Forest Service, Pacific Northwest Research Station. Available from http://www.fs.fed.us/pnw/pubs/pnw_gtr923.pdf? (accessed December 10, 2015).

Wagner, S.R. (2001) *Sisters in spirit: Haudenosaunee (Iroquois) influence on early American feminists.* Summertown, TN: Native Voices Books.

Weaver, H. (2009) The colonial context of violence: Reflections on violence in the lives of Native American women. *Journal of Interpersonal Violence*, 24(9), pp. 1552–1563.

Whyte, K.P. (2014) Indigenous women, climate change impacts, and collective action. *Hypatia*, 29(3), pp. 599–616.

World Health Organization (WHO) (2011) Gender, climate change and health. Available from http://www.who.int/globalchange/publications/reports/gender_climate_change/en/ (accessed August 8, 2013).

13 Youth creating disaster recovery and resilience in Canada and the United States

Dimensions of the male youth experience

Jennifer Tobin-Gurley, Robin Cox, Lori Peek, Kylie Pybus, Dmitriy Maslenitsyn and Cheryl Heykoop

Gender and disaster research has clearly identified the ways historical patterns of gendered power dynamics, household responsibilities and the division of labor have placed women and men in different social locations before, during and after disaster (Alway and Smith, 1998; Enarson 2000; Enarson *et al.*, 2006; Enarson, 2012). Fothergill and Peek (2015) call attention to the fact that these gender differences are often documented in research, but are rarely explored in the context of male and female children and youth in disasters. Because the expectations and experiences of children and youth shift across time and space (Corsaro, 2011), including in disaster settings, it is important to understand their life worlds through a gendered lens and not solely as derived from what is known about adult gendered behavior.

This chapter draws from ongoing research to highlight questions about male youth and the complexity and diversity of their post-disaster experiences in Canada and the United States. We begin by reviewing relevant literature and introducing the Youth Creating Disaster Recovery and Resilience project, a cross-cultural research study that we draw from to support this discussion. We then explore two major realms of social life where gender patterns among youth are clearly differentiated—gender roles and emotional responses to disaster—highlighting the experiences and feelings of boys and young men specifically. Finally, the chapter concludes with ideas for future research.

Gendered youth

For youth in contemporary society, masculinity and femininity are learned and reproduced through family, peer and school interaction (Paechter, 2007). Within these institutions, children and youth are learning what roles are socially acceptable for them to perform. Youth femininity is traditionally characterized by caring behavior and investment in appearance, and youth masculinity by risk-taking behaviors (Paechter, 2007). However, as notions of masculinity and femininity become more fluid, the stereotypical masculine and feminine gender roles are also becoming more nuanced (Nayak and Kehily, 2008). There is an increasing recognition and acceptance of gender and sexuality as existing on a continuum rather than as opposing binaries (McCormack, 2014). Pyne discusses a paradigm shift where the perception of children who do not conform to dominant gender stereotypes has changed "from disorder to diversity, from treatment to affirmation, from pathology to pride, and from cure to community" (2014, p.1). Yet, it is important to note that the growing acceptance of nonconforming gender norms is geographically situated and contextual, and is affected by factors such as socioeconomic status, parental education levels and family structure and size (Priess *et al.*, 2009). In areas where traditional gendered expectations still dominate much of the social discourse, it is important to investigate how these norms shape responses to and expressions of trauma for children and youth. For example, expectations of rigid masculine behavior can have negative consequences for the emotional health of boys. A three-year longitudinal study assessing 504 adolescent boys in the United States found that as boys reported "gender-typed behaviors," their depressive symptoms increased and the quality of their friendships decreased (Gupta *et al.*, 2013).

Youth, gender and disasters

It is important to explore how this gendered "way of being" plays out for boys when their worlds are transformed by disaster. In *Children of Katrina*, Fothergill and Peek (2015) highlight ways that children mirrored adult gender roles, emotions and behaviors. They found that although there were instances of deviation from traditional gender roles following Hurricane Katrina, there were many more cases where children assumed responsibilities similar to that of their same-sex parent. For example, boys frequently assumed "outside" physical work similar to the men, while girls modeled the caretaking roles of their mothers. Fothergill and Peek state, "Not only are boys and girls reflecting the gendered behavior of adults, they are also responding to social cues and expectations about 'appropriate behaviors' for each gender" (2015, pp. 212–213). Psychological and behavioral science research on children and disasters has recognized a pattern where girls tend to express more emotional distress than boys, although more research is needed to establish which boys and girls may be most at risk of negative psychological outcomes and why (Masten, 2011; Masten and Narayan, 2012).

Much of the literature focusing on gendered disaster vulnerability of youth is rooted in the historical inequities experienced by females such as economic instability, increased rates of domestic and sexual violence, lack of access to resources, increased caregiving responsibilities, and domestic labor (UNICEF, 2011). Further research is required to understand the preexisting patterns that may increase the vulnerability of male youth following a disaster. For example, during non-disaster times in the United States, adolescent boys are more likely than girls to drop out of school, receive lower grades, receive disciplinary action, be diagnosed with behavior and mental health problems, commit suicide and homicide, and experience physical violence (Watts and Borders, 2005). Given what is known about the potential for disasters to amplify patterns of vulnerability (Thomas *et al.*, 2013), it is important to understand more specifically how boys' lives may be disrupted by extreme events. For instance, male youth school attendance and performance may be especially problematic in the aftermath of disaster. They also may be more prone to bullying and being bullied in school, to drug and alcohol abuse, and to other risk-taking behavior.

Clearly, there is variability in terms of the vulnerability, capacity and social location of children and youth in disasters. Empirical research on disaster impact and recovery that is both gender and age sensitive is rare, especially in high-income societies. However, in the United States, male youth mortality from "forces of nature" (including natural disasters) was found to exceed that of female youth across the 0–24 age groups studied (Zahran *et al.*, 2008). In a 2008 survey of high school students in Texas following Hurricane Ike, boys were found to be significantly less likely than girls to evacuate (Temple *et al.*, 2011). The nonevacuating boys were significantly more likely to report being sexually assaulted by their dating partners than were their evacuating male counterparts; they also reported higher levels of perpetrating both physical and sexual teen dating violence as well as substance abuse (Temple *et al.*, 2011). In this same context, African-American teenage males whose residences were damaged were twice as likely to report they had abused their intimate partner in the aftermath (Meshack *et al.*, 2012). With respect to climate change hazards and adaptation, boys are somewhat more visible. In a study of child-centered disaster reduction conducted in Indonesia (Haynes *et al.*, 2010), there were clear differences in how boys perceived risk and in their judgments of the capacities of girls of their age. These findings alone warrant far more investigation into the lives of boys following disaster.

Youth creating disaster recovery and resilience

Throughout the remainder of this chapter, we discuss the post-disaster experiences of youth (ages 13–22) in Canada and the United States. We recognize these experiences may vary dramatically from the experiences of youth in other parts of the developing and developed world, and as such this limits our ability to make broad conclusions about the gendered experiences of youth in disaster. Yet, ignoring these distinctions has too long been a limitation in discussing the realities of young people, and our insights offer a starting place to consider the experiences of male youth in disaster contexts.

The Youth Creating Disaster Recovery and Resilience (YCDR[2]) project (ycdr. org) sought to provide a space for young survivors to share their disaster experiences. While we did not place gender at the center of our research agenda or analysis, it is obviously central to the ways individuals experience and interpret their social worlds. As such, our data offer early insights into how male youth navigate post-disaster settings and how these experiences are, indeed, gendered and warrant further consideration. The YCDR[2] research team led creative workshops, interviews and/or focus groups with 80 youth (age 13–22) and 112 adults in six Canadian communities (Bowness, Calgary, Canmore, High River, Morley, Slave Lake) and one in the United States (Joplin, Missouri). This chapter draws primarily on data gathered from workshops that included 10 male and 18 female youth in High River and Slave Lake, Alberta, Canada and in Joplin, Missouri in the United States.

Findings from the YCDR² Project

Gender Roles

Complementing findings presented in *Children of Katrina* (Fothergill and Peek, 2015), male youth in the YCDR² study assumed both traditional and nontraditional gender roles following disaster. However, the females represented in our study did not deviate from the typical patterns of caretaking that we often see among women in the post-disaster context. We heard many stories from male youth in Canada and the United States that were both consistent with the typical gender roles of men, such as engaging in physical labor, as well as stories about them stepping into caretaking alongside their female peers.

As expected, male youth often described participating in outside work activities that involved physically demanding or even risky roles. For example, one 16-year-old male was tasked with assisting his father (a medical doctor) in running a triage station in the immediate aftermath of the Joplin tornado; others described how they were involved in debris removal. One 18-year-old male explained how he got to help rebuild homes in his community following the 2011 Slave Lake wildfires:

> For me, I got to see it different 'cause I actually worked with my builder building five new houses … I went through probably about half the stages required for each of the houses and I got to see the final products for some of the ones I missed out on. Like, I laid a foundation for the final house.

And after the Joplin tornado another male youth similarly explained,

> These [housing contractor] guys actually really helped me during the tornado. They taught me a lot of things about construction stuff. The shed, it was okay, but we needed a new roof and I helped with that. We had to re-shingle the top half. It was pretty cool … I think my mom asked them, because my mom wanted me to be busy with other things so I would not have time to think.

In contrast, we also heard stories from youth that deviated from gendered expectations of men in disaster. For example, during a group interview, one teenage male described stepping into a caretaking role with his younger siblings after the 2011 Joplin tornado. He explained:

> We spent a lot of time with younger children. Adults tended to stay more with other adults. And I don't know if any of you guys experienced that too? But we were kind of left with ourselves. So, having to depend just on adults. You know, there were kids younger than us in the same position and so we stepped up and set an example, and just try and guide them.

This was echoed by another male youth in Joplin who described taking on a mentorship role with younger children in order to fill the gap left by parents who were busy managing the physical and practical elements of recovery:

We were true friends to them [younger children], because while their parents were out trying to get stuff, we were just there to be their friend and help mentor them with their issues. Churches would have the kid area where they would babysit, but I mean, it wouldn't be adults, or it would be one adult who wasn't doing much because they were all worried about their house. So, it would be the teens in there giving their time where they could be helping, just staying in the church and giving their time.

While male youth in our study adopted traditional and nontraditional gendered roles following the disaster, the girls and young women often described their involvement in caring for children, helping their parents fill out paperwork, and volunteering with their churches, and none of them reported being involved in more physically demanding or risky responsibilities such as debris removal or rebuilding.

Emotional disaster responses

Traditional gender norms typically characterize girls as emotional, caring individuals, whereas boys are expected to be stronger and less emotional. Kivel (2013) explains that while boys experience the same range and complexity of feelings as girls, they often come to suppress many of their emotions. For example, boys are more likely to be reprimanded for crying or conveying sadness, while the same behavior is often supported and encouraged for girls (Kane, 2006; Peek and Fothergill, 2008; Fothergill and Peek, 2015). This gendered differentiation between the encouraged emotional responses of boys and girls was evidenced in our research. For example, a male youth explained that it was his job to be strong and to let his mother grieve after they endured the Slave Lake wildfire. He recalled, "My mom had a pretty difficult time with it [the fire] but it was pretty much mine and my dad's job to make sure that she made it through." In contrast, one female youth in our study explained the support of her mother when she returned home for the first time following the 2011 Slave Lake wildfires:

I didn't know what reaction I would really have to it. Even my mom too, she was like "it's okay if you cry when you see it." And it's just … like it was crazy to see for the first time and it's just unbelievable really.

Kivel (2013) notes above that boys experience a similar range of emotions as girls; but, as Rose *et al.* (2012) point out, boys and girls express emotions differently and for different reasons. In their study of youth friendship, they found that, while girls may disclose personal information to friends because they anticipate positive emotional outcomes, boys indicated that sharing was a waste of time, made them feel weird and did not yield any perceived benefits. Emotional expression is a socially constructed and highly gendered process (Fivush and Buckner, 2000). Children learn from an early age that sadness and emotions should be communicated in culturally appropriate ways and this becomes even more concrete in adolescence (Fivush and Buckner, 2000). Although feelings of rage and anger are often expected and tolerated in male culture, one of the strongest gender stereotypes is that of the stoic man—and the crying woman.

Following a disaster, some researchers report that boys tend to "externalize" their feelings through negative behaviors such as violence and drug use while girls "internalize" their emotions, which can lead to depression and anxiety (Masten and Narayan, 2012; Fothergill and Peek, 2015). While this may be true, male youth in our study described outlets that included service to others (e.g. debris removal or helping at the church) and creative expression as ways of processing their feelings. One male confirmed that instead of talking to others, playing music after the 2011 Joplin tornado was more helpful in his process of dealing with the trauma of the disaster. He explained that playing the guitar:

> gave me an outlet to express myself and I think it does for a lot of musicians; and the same with painters or dancers or anything. That is their way of expressing how they feel. Cause a lot of the time people can't express themselves just by talking.

The workshops for our YCDR[2] study were not designed to specifically assess emotional coping strategies, but during the course of the study male youth regularly discussed external forms of release such as playing video games, as this young man from High River explains after the 2013 Alberta floods:

> It is force of nature. Suddenly, your escape from reality is no longer a possibility. The only thing you can think about is the scary thoughts; like a broken home, the stress that your parents are feeling, of everything that you have lost. I know that at least for me, having my PS3 [Play Station 3 video game console] with my favorite game saved my soul from all of the negativity.

Two other male youth, both Joplin tornado survivors, described how they channeled their negative emotions through writing. Both boys said this helped them with the emotions they were feeling and also to get the "anger out" in a positive way, rather than "hurting somebody." Another teen in High River told us that he started telling jokes and "cutting up" in silly ways to process the pain and loss after the 2013 Alberta floods.

Male youth also shared how important engaging in physical activity was to help them process their disaster experiences. One male teen from Joplin focused his energy on martial arts, while another described how he processed the internalized stress that he felt following the 2011 tornado:

> Running is one of my major coping strategies. About a year after the tornado, the first couple months, it was eating. I gained eighty pounds, and then I lost it all from running. So I got up to 285 [pounds] at one point and definitely running has helped me.

For the most part, the boys in this study did not emphasize relationships with friends and family as central to their post-disaster coping processes. This was in direct contrast to the many stories that the girls and young women in the study shared regarding the importance of these relationships. Boys and girls also differed in *how* they shared their feelings and ideas. This was reflected in the following exchange among four youth in Slave Lake, all of whom were affected by the wildfires that destroyed nearly one-third of their community:

Female 1:   I think it's girls that are more open to talking about our emotions.
Female 2:   Yeah.
Female 1:   As you can tell, we kinda dominated the conversation.
Male 1:   Hey!
Male 2:   We just got pushed out, hey Doug?
Male 2:   We should just be like "beer buddy?"
Male 1:   Yeah.

The females in this conversation suggest they are more able and willing to discuss their emotions and that, as a result, have "dominated" the conversation. The males acknowledge this dynamic, but suggest a stereotypically alternative male space—conversation over beer.

New research directions

Drawing from the existing literature on youth and disasters and the preliminary findings from the YCDR[2] project, it is clear that additional research is needed to examine the gendered dimensions of youth and disaster recovery. As such, we have identified several areas for further research. First, we recommend that future studies observe the gendered roles and responsibilities assumed by youth in the aftermath of a disaster and investigate how these influence the recovery process. Are youth who are directly involved in recovery activities more likely to achieve emotional and physical recovery sooner than their peers who are denied the same opportunities? How and why do girls and boys tend to get involved in recovery activities? How are the actions of young people shaped by adult expectations, requests for assistance and limits placed on the behaviors and actions of youth? Does adhering to traditional gender roles help boys and girls or does it hinder recovery?

Second, we suggest a closer examination of emotional expression and the coping mechanisms male and female youth draw upon in order to assist mental health professionals to offer more responsive services post-disaster. Are only females being encouraged to cry and grieve following disaster, and are males sanctioned for similar behavior? What role do female and male youth play in creating and maintaining emotional space after a disaster? Are boys and men expected to process pain and emotion after a disaster in ways that are outside of their comfort zone? Do these patterns shift in ways that affirm or contradict the dominant norms constraining young men and boys? If so, for how long and why? Is there more latitude for nonconforming gender expression for male and female youth in crisis contexts—or less? How do boys' relationships with one another and adult men change, if at all, and how does this matter? In addition to the aforementioned questions, we think it is important not just to explore how youth cope in terms of their internalizing or externalizing behaviors, for instance, but also to carefully study whether they ask for help at all and who they turn to for help when they do need some form of support. Of course, these areas should be examined through an intersectional lens, to better understand the ways that gender, race, class, ability and other characteristics may influence these help-seeking behaviors.

Third, we suggest further that additional research be conducted to clarify *when* and *how* youth feel comfortable sharing their disaster experiences and if this does in fact result in positive outcomes. Do boys and girls prefer to share in one-on-one or group settings? Does sharing alleviate or add stress to young people's lives, based on differential gendered expectations regarding the ability to share?

Fourth, amidst the changing landscape of the gender dichotomy, we recommend that gender identity and gender expression be considered in future research to gather a more holistic understanding of the gendered dimensions of youth and disaster recovery. While it is clear that youth often do assume similar gendered roles as has been documented among adults in the aftermath of disaster, there are also spaces and times when these traditional roles and relationships are disrupted. More careful examination is obviously in order.

Conclusion

Through the voices of youth participants in the YCDR² study, we illustrate the ways that male youth may follow and deviate from traditional gendered roles, emotional expression and coping behaviors. It is clear that gendered differences noticed in adults affected by disasters are in many ways mirrored in the perceptions, behaviors and comments of youth in this study. Just as researchers have concluded that women and men typically adhere to traditional gender roles after a disaster (Enarson, 2012), the findings suggest that this holds true for youth as well. Children and youth, like women and men, are not a unified or homogenous group of people. Although our findings did not explore this complexity, it is important to acknowledge that these youth also differ by race, class, social location, developmental capacity, sexuality and other significant identities. It is important that researchers use robust qualitative and quantitative methods to begin to examine differences between and among groups of children and youth, as well as between girls and boys across a range of social determinants, to uncover the harmful and protective factors that may emerge for persons of all ages and genders in the aftermath of disasters.

References

Alway, J. and Smith K. (1998) Back to normal: Gender and disaster. *Symbolic Interaction*, 21(2), pp. 175–195.

Corsaro, W.A. (2011) *The sociology of childhood*. 3rd edn. Thousand Oaks, CA: Pine Forge Press.

Enarson, E. (2000) "We will make meaning out of this": Women's cultural responses to the Red River Valley Flood. *International Journal of Mass Emergencies and Disasters*, 18(1), pp. 39–62.

Enarson, E. (2012) *Women confronting natural disasters: From vulnerability to resilience.* Boulder, CO: Lynne Rienner Publishers.

Enarson, E., Fothergill, A. and Peek, L. (2006) Gender and disaster: Foundations and directions. In: Rodriguez, H., Quarantelli, E.L. and Dynes, R.R. (eds.) *Handbook of disaster research.* New York: Springer, pp. 130–146.

Fivush, R. and Buckner, J.P. (2000) Gender, sadness, and depression: The development of emotional focus through gendered discourse. In: Fischer, A. (ed.) *Gender and emotion: Social psychological perspectives.* Cambridge: Cambridge University Press, pp. 232–253.

Fothergill, A. and Peek L. (2015) *Children of Katrina.* Austin: University of Texas Press.

Gupta, T., Way, N., McGill, R.K., Hughes, D., Santos, C., Jia, Y., Yoshikawa, H., Chen, X. and Deng, H. (2013) Gender-typed behaviors in friendships and well-being: A cross-cultural study of Chinese and American boys. *Journal of Research on Adolescence,* 23(1), pp. 57–68.

Haynes, K., Lassa, J. and Towers, B. (2010) Child centred disaster risk reduction and climate change adaptation: Roles of gender and culture in Indonesia. Working paper No. 2 (September). Available from http://resourcecentre.savethechildren.se/sites/default/files/documents/3960.pdf (accessed November 10, 2015).

Kane, E.W. (2006) "No way my boys are going to be like that!" Parents' responses to children's gender nonconformity. *Gender & Society,* 20(2), pp. 149–176.

Kivel, P. (2013) The acts like a man box. In: Kimmel, M.S. and Messner, M.A. (eds.) *Men's lives,* 9th edn. Upper Saddle River, NJ: Pearson Education, pp. 14–16.

Masten, A. (2011) Resilience in children threatened by extreme adversity: Frameworks for research, practice, and translational synergy. *Development and Psychopathology,* 23, pp. 493–506.

Masten, A.S. and Narayan, A.J. (2012) Child development in the context of disaster, war, and terrorism: Pathways of risk and resilience. *Annual Review of Psychology,* 63, pp. 227–257.

McCormack, M. (2014) The intersection of youth masculinities, decreasing homophobia and class: An ethnography. *The British Journal of Sociology,* 65(1), pp. 130–149.

Meshack, A., Peters, R., Amos, C., Johnson, R., Hill, M. and Essien, J. (2012) The relationship between Hurricane Ike residency damage or destruction and intimate partner violence among African American male youth. *Texas Public Health Journal,* 64(4), pp. 30–33.

Nayak, A. and Kehily, M.J. (2008) *Gender, youth and culture: Young masculinities and femininities.* London: Palgrave Macmillan.

Paechter, C. (2007) *Being boys, being girls: Learning masculinities and femininities.* New York: Open University Press, McGraw-Hill.

Peek, L. and Fothergill, A. (2008) Displacement, gender, and the challenges of parenting after Hurricane Katrina. *National Women's Studies Association Journal,* 20(3), pp. 69–105.

Priess, H., Lindberg, S. and Hyde, J. (2009) Adolescent gender-role identity and mental health: Gender intensification revisited. *Child Development,* 80(5), pp. 1531–1544.

Pyne, J. (2014) Gender independent kids: A paradigm shift in approaches to gender nonconforming children. *Canadian Journal of Human Sexuality,* 23(1), pp. 1–8.

Rose, A.J., Schwartz-Mette, R.A., Smith, R.L., Asher, S.R., Swenson, L.P., Carlson, W. and Waller, E.M. (2012) How girls and boys expect disclosure about problems will make them feel: Implications for friendships. *Child Development,* 83(3), pp. 844–863.

Temple, J.R., van den Berg, P., Thomas J.F., Northcutt, J., Thomas, C. and Freeman, D.H. (2011) Teen dating violence and substance use following a natural disaster: Does evacuation status matter? *American Journal of Disaster Medicine*, 6(4), pp. 201–206.

Thomas, D. Phillips, B., Lovekamp, W. and Fothergill, A. (eds.) (2013) *Social vulnerability to disasters*. 2nd edn. Boca Raton, FL: CRC Press.

UNICEF (2011) *The State of the World's Children 2011. Adolescence: An age of opportunity*. New York: United Nations Children's Fund.

Watts, R.H. Jr. and Borders, D. (2005) Boys' perception of the male role: Understanding gender role conflict in adolescent males. *The Journal of Men's Studies*, 13(2), pp. 267–280.

Zahran, S., Peek, L. and Brody, S.D. (2008) Youth mortality by forces of nature. *Children Youth and Environments*, 18(1), pp. 371–388.

Part IV
Transforming masculinity in disaster management

14 Firefighters, technology and masculinity in the micro-management of disasters

Swedish examples

Mathias Ericson and Ulf Mellström

This chapter addresses how the strong link between technology and masculinity is forming the occupational ethos of firefighters and how the micro-management of disasters in operational service is directly connected to this gendered work practice. The celebrated masculine ethos of the firefighting profession is formed through an intricate process of embodied practical skills, technical dexterity and a capacity to act. As such, it carries a particular symbolic weight of classic masculine heroism and is also one reason why firemen have been so celebrated historically. The heroic men preventing disasters and accidents and rescuing people are also connected to a wider cultural imaginary of communal unity, protection and safety, where their form of masculinity epitomizes a cultural ideal of a particular good-hearted masculine heroism.

However, this specific heroism is also an explanatory key for why women are directly and indirectly excluded from the fire services. The masculine occupational heroism of firefighters is widely and commonly honored and valued, but also makes the profession resistant to critical inquiry with regard to gender equality, as well as other affirmative stances connected to sexuality and race (Baigent, 2001; Ward and Winstanley, 2003, 2005, 2006; Ward, 2008; Olofsson, 2009; Ericson, 2014). This hallmark is not easily questioned, and is also key to understanding slow progress in including female firefighters in the Swedish rescue services, as well as in the rescue services in many other countries (Baigent, 2001). Today, the percentage of female firefighters is just above 3% in the Swedish rescue services. This still modest representation of women is equivalent to the number of female firefighters in other countries such as Japan, US, Germany, Canada, Australia and UK.

Resistance to including women in the rescue services is still omnipresent, although few firefighters would claim being positioned in this way in a context where gender equality is regarded as a shared cultural value (Ericson, 2010). The latter is to be understood in the wider context of the Swedish welfare state, and where a strong gender equality discourse is a prominent theme in the history of the Scandinavian countries, particularly Norway and Sweden (Holth and Mellström, 2011). Among other things, this implies a contemporary political discourse around the affirmative values of gender equality shared by seven out of eight political parties in the Swedish parliament.

In the institutional setting of the Swedish rescue services, we can observe how such a discourse in the last 10 to 15 years is connected to new legislative measures that emphasize proactive work and the prevention of accidents, in combination with efforts to gender mainstream work processes and strengthen work on gender and diversity issues in the rescue services. We see increasing interest in how to implement gender mainstreaming and to find tools for proactive gender equality measures. This proactive approach, together with the prescribed goal of gender and diversity management in the rescue services, includes increasing the representation of women and men from various backgrounds and with different expertise in emergency management.

It is against such a background of masculine occupational heroism and a strong prevailing gender equality discourse that we investigate in this chapter how concrete practices in the micro-management of crisis and disaster work are guided by normative structures that exclude most women and many men of the "wrong kind." We also show how certain kinds of technology are preferred because of their connection to the idealized form of a man of action. In particular, we trace how processes of resistance are connected to technical know-how, artifacts, bodily dispositions and safety regulations procedures. However, we also trace a high degree of ambivalence toward the introduction of new technologies as this new equipment often carries a distributed and open-ended agency. By open-ended agency, we mean that the technology can be used by various end-users, and not exclusively by professional firemen.

The empirical insights of this chapter draw on long-term research in the Swedish rescue services (Ericson, 2004, 2010, 2011, 2014), with a special focus on masculinity constructions, rescue services and firefighters. The empirical material includes observations, interviews and interactive seminars with firefighters on issues such as gender equality, proactive work and the symbolic value of masculinity. The material was collected in two different periods. In 2002–2005, Ericson did fieldwork studying firefighters' everyday life at the station and operations on the sites of accidents, focusing on masculinity construction and homosociality. In 2010–2013, the authors collaborated in a follow-up study that investigated such proactive work as building trustworthy relations in socially marginalized suburbs, or demonstrating safety procedures at nursing homes. We were interested in how this might challenge masculinity constructions and homosocial practices within this profession. Altogether, the material includes interviews with 30 firefighters and 10 fire chiefs and/ or fire engineers at 14 different fire stations. The interview age span ranged from 23 and 58 years, although the second study focused more specifically on younger firefighters. Over two-thirds of those included are white men.

We begin below by outlining the theoretical points of departure that connect masculinity, technology, embodiment and heroism with regard to firefighters and the rescue services.

Firefighters, heroism and masculinity

As noted, there are few, if any, occupations that carry the same particular symbolic weight of classic masculine heroism, and few that have been so celebrated historically as firemen in bringing out the "interconnecting qualities of ideal manhood to which he was assimilated: manliness/masculinity, chivalry and heroism" (Cooper, 1995, p. 141). This heroism also extends to what Whitehead sees as a constitutive feature in almost all forms of masculinity: "For most men, any 'heroic project' begins when they leave for work" (2002, p. 123). The conflation of heroism, masculinity and firefighting is also traditionally related to class, where the occupational skills of firefighters are closely connected to an ethos of white working-class masculinity (Baigent, 2001; Desmond, 2007; Carroll, 2011).

As Faludi (2008) and others have pointed out (Baigent, 2001; Lorber, 2002; Chetkovich, 2004; Tracy and Scott, 2006; Carroll, 2011), these intersecting qualities are also commonly associated with a deeply rooted metaphor of masculine achievement as a cultural imaginary of national unity and protection, not least symbolized by the firefighters who rescued people from the Twin Towers of New York City on September 9, 2001. The heavily loaded symbolism of such gendered representational practices is what gives the occupation its prominent aura and elevated status, and is also what points to the inertia of change.

Sites of resistance in regard to change in a highly gender-imbalanced organization can usually be identified at various organizational locations. According to Dekker (2007), rescue service organizations are generally slow to adapt to organizational change due to a long tradition of hierarchical structures. If there is a key moment at which resistance against any gendered change in the rescue services can be seen, and where the core of firefighters' masculine heroism resides, it is entering buildings that are on fire and conducting smoke diving operations. Several researchers (Chetkovich, 1997; Baigent, 2001; Lorber, 2002; Tracy and Scott, 2006) have shown how this situation is regarded as the ultimate test of trustworthiness among male colleagues.

The homosocial processes identified in the fire service parallel what have been observed in other male-dominated professions such as engineering or typesetting (Cockburn, 1991; Mellström, 1995; Tracy and Scott, 2006). Similar arguments resting on gender complementarity and essentializing discourses are here used to protect certain forms of exclusively male homosociality. The balancing point of such discourses is often recognized in masculine heroic images, celebrating courageous deeds and bravery mythologized in stories of occupational heroes, all retold and mediated through collective remembering practices that exemplify the core values of the occupation (Mellström, 1995).

The mobilization of homosocial emotions within classical masculine occupations such as the rescue services is also often connected to technologies and artifacts in various ways. Such practices of masculine fraternity, machines and technology can often be understood as a means of embodied communication enabling homosocial bonding between peers as well as between generations of men. In other words, in the construction of masculine fraternity, tools and artifacts become an essential element in the sharing of these relationships, and are often part of what it means to be a man, part of a masculine script. Technology and the mastering of tools have often been an essential part of many men's upbringing as boys, and connect closely to definitions of what is masculine and what is not (Mellström, 2002, 2003, 2004).

An open-ended agency

Not surprisingly, we also observe that dexterity in the mastering of different tools and machinery is vital to the capacity to act and essential to the professional ethos, closely intertwined with a sense of masculine pride. Tools make it possible for firefighters to be positioned at the center of disasters such as devastating bush-fires, terror attacks or burning buildings and ferries, and to do things that other-wise would be impossible. Secondly, tools are essential not only to the safety of firefighters but also to manifesting their position as life savers for ordinary citi-zens. The importance of acquiring embodied practical skills in close interaction with different tools is illustrated by many researchers.

Olofsson (2009) and Weick (1993, 1996) provide different reiterations of two of the most historically well-known disaster scenarios: South Canyon, Colorado (1994) and the Mann Gulch, Montana wildfires in 1949 in which 23 firefighters (19 men and four women) died. They propose that these cases could teach us about how key organizational and individual vulnerability are to understanding the conceptual triad of tools, firefighting and gender. While Weick (1993) does not employ a gender perspective, Olofsson builds on his analysis by adding a gen-der perspective that puts the disastrous accidents in a slightly different light. The accidents of Mann Gulch and South Canyon are particularly noteworthy because the dead firefighters failed or refused to drop their tools when retreating from the raging flames. Weick (1993) lists ten reasons to explain why the firefighters would not let go of their tools, and Olofsson brings four of these to the fore to help understand the lingering question of why they failed to drop their tools: control, identity, skill at dropping and failure.

Olofsson's key to understanding the fatal consequences of Mann Gulch and South Canyon is to weave together masculinity, tools and homosocial practices in male-dominated workplaces as negotiated and enacted in close collaboration in a heterogeneous network of human and nonhuman actors. In short, Olofsson's approach extends agency to a network of interconnected processes merging the social and the technological. In line with this, we see the close connection between technology and masculinity as fundamental to the professional ethos of firemen who act as men within the heroic imaginary of their occupation. However, we also witness their ambivalent relation to various forms of technologies and machinery, for these carry an open-endedness, or distributed agency (Hutchins, 1995), that challenges the occupational ethos of a man of action.

Resistance, masculine ideals and technology

As we have argued thus far, we see the process of embodying of tools and different artifacts as essential to firefighters' professional ethos (saving lives as physically tough men of action) but with an ambivalent relation to technology. This is some-thing we observe in particular with regard to the introduction of new technology. For instance, Baigent (2001) and Chetkovich (1997) provide examples of a nostalgic glorification of the "leather-lunged" old timers, who did not bother to wear a breath-ing apparatus to enter burning buildings in smoke diving operations. In a similar vein, Olofssons' (2011) study of retired firefighters shows that the introduction of new technology was considered a feminization of masculine heroism in the professional ethos.

The profession of firefighting has traditionally been closely related to such other male-dominated professions as truck drivers, carpenters, electricians, car mechanics and construction workers. Firemen have most often been recruited from these professions. Following from this, a preference for mechanical know-how has been highly valued, although always in the service of the professional mission of saving lives. These two components constitute a baseline in our interviews and observations. On the one hand, the firemen are truly dependent upon their embodied technical skills and modes of action. On the other hand, they repeatedly commented that their tools have to be simple and easy to handle. To exemplify, an infrared camera used in smoke diving operations was demonstrated by one fireman during Ericson's fieldwork, with the comment that it was "firefighter adapted" because, as the interviewed fireman said, "it's just a single button" technology.

Work in the rescue services requires a constant engagement with different forms of technology such as carrying baskets of hoses or cutting tools, climbing up ladders or entering smoke-filled environments using breathing apparatus. It is also these practices that form the core of the professional identity and dominate in the public eye. But we could also observe that firefighters were cautious about giving the tools too much credit, pointing out that the tools in themselves where rather simple. The important thing was knowing how and when to make the best use of them. For instance, full-time employed firefighters would ridicule part-time firefighters as simple posers who lacked any "real" experience in accidents (Ericson, 2011). Occupational bravado is negotiated in the interstices between being men of action, and men who carry the embodied skills of essential tools and artifacts that do the job effectively. These technical skills must furthermore conform to homosocial norms that serve to distinguish reliable (good) teammates from individualists and free riders.

The balance between being a good teammate and harboring the necessary embodied technical skills also connects to the skepticism and resistance that we found was expressed toward increasingly formal training. We will exemplify how this resistance is articulated in daily work practices with one episode from Ericson's fieldwork:

> On one occasion, two of the younger firefighters disassembled the breathing apparatus in the maintenance room in order for it to be serviced at the fire station. They asked if I would like to join them and explained that I was allowed to disassemble a breathing apparatus if I wanted to, but that you needed documented formal training to put the breathing apparatus back together. After a while, one of the senior firemen joined us, placing himself at the table where some of the still disassembled breathing apparatus were placed. As he listened to the discussion, he picked up the parts of a breathing mask and started to quickly put it back together. He seemed so accustomed to the practice that he did not even have to look at what he was doing. It did not take long before the young firemen ended their discussion. The senior fireman then

started commenting sarcastically that he did not know how to put a breathing apparatus together anymore because he did not have the proper training. It was obvious that he had performed the procedure many times and displayed a truly embodied competence of how to service a breathing apparatus. After he finished, he took one of the stickers that was to be placed on the breathing apparatus to signal that it was done and waved it in the air: "Where should I put this for instance? Could I put it here? I do not know because I don't have the proper training." He then left the room.

Normally, the required formal training in maintenance and repair of technical equipment is introduced as an intentional improvement, protective of firefighters' safety and making routine procedures more transparent. However, this was not necessarily acknowledged by all firefighters, as seen in the excerpt above. Maintaining and servicing one another's breathing apparatus used to be a task done, and collectively, in the teams. However, this routine procedure has recently been formalized, now requiring formal education.

This formalization of technological know-how has challenged the traditional hierarchy within work teams, since working procedures were previously based on the number of years one had served as a fireman. This meant that young firefighters would be assigned to service the breathing apparatus, while senior firemen could not formally engage in these work practices. The tension that the incident above created is associated with age as well as gender, and is directly connected to the formalization of formerly tacit knowledge and it is central to the dominant masculine ethos in firefighting. With respect to reactive/proactive processes of change currently underway in the rescue services, we now witness an organizational struggle materialized through formalization of work practices, and an open-ended agency around the technology that challenges institutionalized masculinity.

Technological change challenging the action-oriented occupational ethos

New technology is continuously developed in the rescue services with the overall aim of distributing surveillance responsibility and response, and an anticipated reduction of risk. This could, for instance, be new information technology and radio communication systems, portable sprinklers and fire extinguishers. Beyond making professional firefighting safer, new technologies are seen to develop resilience generally by teaching other professionals to become first responders, and by empowering ordinary citizens with smoke alarms, fire extinguishers and portable sprinklers.

Firemen are not the only agents in responding to accidents and preventing smaller accidents from escalating to disasters. For instance, the response of care workers is crucial and has gained a more central position thanks to the development of smoke alarms and fire extinguishing techniques. Caregivers are now equipped with fire extinguishers that can simply be thrown into the fire to effectively extinguish a fire long before firefighters can arrive on the scene. In such cases, the technological development obviously challenges the privileged action-oriented ethos of firefighters, in line with a broader proactive approach where it is stressed that fire safety and protection needs should be viewed as a joint accomplishment.

This technological development is a cause of anxiety among firefighters in the sense that they become increasingly disconnected from risk and risk-taking, now more distanced from hands-on life-saving operations. New technological equipment adapted to improving operational efficiency, such as tools that make it possible to extinguish fires without having to actually enter the building, or new extinguishers that cut through walls with high pressure water and extinguish fires by spreading a fog of water. These improved ways of extinguishing fire reduce dependency on individual smoke diving, indirectly challenging the long-lived professional firefighting ethos based on brave and heroic men of action.

Parallel to the introduction of such new technologies, structural changes have pointed to a more proactive approach in the Swedish rescue services since the mid-2000s. In this context, "proactive" refers to early intervention and prevention in a risk scenario. New legislation and reformed education for firefighters put more focus on proactive work and on empowering local communities to reduce fire risk and respond to fire. However, these changes have met considerable resistance. It is generally argued that these undermine the traditional and valued reactive capacity of the rescue services. In an interview, one watch manager said:

> We don't have time to practice anymore, it is ridiculous. Just have a look at all the tools and equipment that is placed on the firetruck. The truck is basically just a really big toolbox. And we need to practice so that we know how to use every single one of them.

In our research, we observed a constant tension and negotiation between proactive and reactive agendas. One interviewed firefighter described a "proactive" project to us, one based on the idea that it should be possible to divide teams of firefighters into smaller work units. The minimum safety standard for smoke diving is now one watch manager, two smoke divers and two back-up smoke divers. However, for most firefighters, smoke diving is an operation performed very seldomly. It has been suggested, therefore, that it should be possible for firefighters to take on more proactive work such as demonstrating basic fire prevention to school children or care workers. It would thus be possible for firefighters to work in teams of fewer than five persons between calls, and especially at fire stations with low call frequencies.

Another response is the development of smaller vehicles carrying some basic equipment (e.g. the cutting extinguisher) so that even a small unit could start up work at accident sites until joined by the full team. One firefighter expressed that he had enjoyed working in this manner, since it made it possible to move around in the city and do proactive work without losing sight of operational work. In fact, the smaller vehicle had a positive effect, as the smaller team arrived first at sites where minutes were precious, such as car crashes and burning houses. Yet, other members of his team ridiculed and criticized the use of the small vehicle, referring to these firefighters as "fake" and "phony," and to these new trends as a "Mickey

Mouse" form of workmanship. After some months of further ridicule, the smaller vehicle was damaged during training and sent to the repair shop. After the vehicle had been repaired, not one man dared to put it to use again due to heated conflicts in the work teams.

In the action-oriented world of firefighting, the collective spirit is truly important but at risk of being undermined as described above. The performative aspect of firefighting and the visible mastering of tools are closely tied to an orientation toward action and reaction, rather than to proaction (in this case, interventions that reduce the risk of harm due to fire). As sometimes expressed by firemen, it is not always what they do that is important, but that they do something. One explained:

> You should never get stuck when doing operations, or get bewildered. You have to do something, no matter what. When people look at you, you have got to be doing something, you must never end up just standing there helpless. You might not be doing the right things, but that is irrelevant as long as you do something. I think that is most important.

The gendered mastery of firefighting tools is something done in teamwork, and often performed in front of an audience gathered at the scene of the accident. This public visibility is an important matter since public legitimacy is at stake in these specific accidents and disasters. It is those moments that ultimately reproduce the heroic male imaginary and symbolic functions of securing trust and security in the local community and beyond. The importance of visibility and public performance, in line with a public cultural script, should not be underestimated. It is part of the intertwined process of embodied practical skills, dexterity and masculinity. Resistance to change in this masculine ethos may also include refusal to use certain tools and machinery, even when these may improve overall efficiency and security for firefighters themselves.

Conclusions

Masculinity is institutionalized through networks of artifacts, representations and work practices that are interwoven with and expand upon certain male bodies. These work practices and institutionalized forms of masculinity long excluded women and "other" men. Tools and machinery have traditionally been effective instruments of exclusion as well as inclusion, and help men secure a valued professional ethos based on masculine heroism and the male embodiment of tools and technologies. Only (certain) male bodies are seen as capable of handling the tools and machinery of effective firefighting.

In the dominance of a new and proactive orientation in the Swedish rescue services, we observe processes that challenge institutionalized forms of masculinity. First, new technologies are being introduced which aim to widen crisis management to include community response and increasing social responsibilities for societal surveillance. We argue that the technologies being introduced have an open-ended agency which can and does empower other professional groups. This is something which challenges the exclusivity of firemen, and in the long run possibly also the cultural imaginary of masculine heroism. Second, as new technologies are introduced to diversify work assignments and enhance operational security in the micro-management of disasters and accidents, fire-fighting becomes more dependent upon complex technologies; as complexity of tools and machinery increases, less room remains for demonstrating collective and individual heroism. Third, as organizations increasingly reorient away from reactive or response-oriented modes of operation, training and recruitment procedures become increasingly formalized. In this way, the classic masculine ethos based in the male working class is challenged, for more proactive approaches to firefighting empower a wider and more diversified professional base, and may engage action-oriented firefighting men in community-based proactive work geared to prevention.

We have shown above how the intricate relationship between technology, masculinity and firefighting is co-constituted. We argue that this conceptual triad is key to understanding how masculinity has been institutionalized in the fire services—and to understanding the root causes of male resistance to change in firefighting.

References

Baigent, D. (2001) Gender relations, masculinities and the fire service: A qualitative study of firefighters' constructions of masculinity during firefighting and in their social relations of work. PhD. Anglia Ruskin University.

Carroll, H. (2011) *Affirmative reaction: New formations of white masculinity*. Durham, NC: Duke University Press.

Chetkovich, C. (1997) *Real heat: Gender and race in the urban fire service*. London: Rutgers University Press.

Chetkovich, C. (2004) Women's agency in a context of oppression: Assessing strategies for personal action and public policy. *Hypatia*, Fall, 19(4), pp. 122–143.

Cockburn, C. (1991) *Brothers: Male dominance and technological change*. London: Pluto Press.

Cooper, R. (1995) The fireman immaculate manhood. *Journal of Popular Culture*, 28(4) (Spring), pp. 139–170.

Dekker, S. (2007) *Just culture: Balancing safety and accountability*. Aldershot, England: Ashgate.

Desmond, M. (2007) *On the fireline: Living and dying with wildland firefighters*. Chicago: University of Chicago Press.

Ericson, M. (2004) *Brandman och man—om aktualisering av kön i brandmannayrket [Firefighter and male: About the actualization of gender in the firefighter profession]*. Karlstad: Räddningsverket.

Ericson, M. (2010) Good manners: Struggles for respectable masculinities and heteronormativities in the Swedish Fire Service. In: Martinsson, L. and Reimers, E. (eds.) *Norm-struggles: Sexualities in contentions*. Newcastle: Cambridge Scholars Publishing, pp. 98–112.

Ericson, M. (2011) Nära inpå: maskulinitet, intimitet och gemenskap i brandmäns arbetslag [Up close: Masculinity, intimacy and community in firefighters' work teams]. PhD. University of Gothenburg.

Ericson, M. (2014) The exceptionalism of firefighters: Heroism, whiteness and masculinity in times of suburban riots. *NORMA: International Journal for Masculinity Studies*, 9(3), pp. 178–190.

Holth, L. and Mellström, U. (2011) Revisiting engineering, masculinity and technology studies: Old structures with new openings. *International Journal of Gender, Science, and Technology*, 3(2), pp. 313–329.

Hutchins, E. 1995. *Cognition in the wild*. Cambridge, MA: MIT Press.

Lorber, J. (2002) Heroes, warriors, and "burquas": A feminist sociologist's reflections on September 11. *Sociological Forum*, 17(3), pp. 377–396.

Mellström, U. (1995) Engineering lives, technology, time and space in a male-centred world. PhD. Linköpings Universitet.

Mellström, U. (2002) Patriarchal machines and masculine embodiment. *Science, Technology & Human Values*, 27(4), pp. 460–478.

Mellström, U. (2003) *Masculinity, power and technology: A Malaysian ethnography*. Aldershot: Ashgate.

Mellström, U. (2004) Machines and masculine subjectivity: Technology as an integral part of men's life experiences. *Men and Masculinities*, 6(4), pp. 368–383.

Olofsson, J. (2009) A momentary lapse of identity and control: Straddling male collaborations with professional instruments. *Journal of Men's Studies*, 17(2), pp. 129–144.

Olofsson, J. (2011) Cut out to handle a ladder? Age, experience and acts of commemoration. *NORMA: Nordic Journal for Masculinity Studies*, 6, pp. 5–20.

Tracy, S.J. and Scott, C. (2006) Sexuality, masculinity and taint management among firefighters and correctional officers: Getting down and dirty with "America's heroes" and the "scum of law enforcement." *Management Communication Quarterly*, 20(1), pp. 6–38.

Ward, J. (2008) *Sexualities, work and organizations: Stories by gay men and women in the workplace at the beginning of the twenty-first century*. London: Routledge.

Ward, J. and Winstanley, D. (2003) The absent presence: Negative space within discourse and the construction of minority sexual identity in the workplace. *Human Relations*, 56(10), pp. 1255–1280.

Ward, J. and Winstanley, D. (2005) Coming out at work: Performativity and the recognition and renegotiation of identity. *The Sociological Review*, 53(3), pp. 447–475.

Ward, J. and Winstanley, D. (2006) Watching the watch: The UK fire service and its impact on sexual minorities in the workplace. *Gender, Work and Organization*, 13(2), pp. 193–219.

Weick, K.E. (1993) The collapse of sense making in organizations: The Mann Gulch disaster. *Administrative Science Quarterly*, 38(4), pp. 628–652.

Weick, K.E. (1996) Drop your tools: An allegory for organizational studies. *Administrative Science Quarterly*, 41(2), pp. 301–313.

Whitehead, S. (2002) *Men and masculinities: Key themes and new directions*. Malden, MA: Polity.

15 Resisting and accommodating the masculinist gender regime in firefighting

An insider view from the United Kingdom

Dave Baigent

As global warming and international terrorism increase the risk of disasters worldwide, it is a considerable concern that first responders to these disasters from the fire service remain predominantly male. This outcome denies the public the benefits of the increase in skills that a mix of gender provides. As the UK's first responder to disasters, firefighters provide multiskilled teams to attend, for example, structural and wildland fires, road traffic collisions, chemical spillages, and floods. However, there are so few women firefighters that the public expect to see men carrying out rescue work at disasters—and male firefighters, for their part, do nothing to challenge what becomes a self-fulfilling prophecy that makes heroes of men (Baigent, 2001). This not only underpins male supremacy, but also puts firefighters in a position of superiority vis-à-vis other masculine groups (Pacholok, 2013).

This gender imbalance is often thought to arise from male hegemony, a taken-for-granted acceptance in which men use their power to make their patriarchal position of dominance and any subsequent dividend seem natural. It is as if it were the view of a whole society—so normalized as to need no explanation (Carrigan *et al.*, 1985; Hearn, 2004, 2012; Hall *et al.*, 2007).

To understand how men are able to have this impact on the difficult terrain of gender in disaster, I am going to use my 30 years of operational experience as a firefighter, and 20 years as an academic and consultant on management, culture and equality to provide a reflexive view of masculinity in the fire service. Using profeminist autocritique (Hearn, 1994), an academically rigorous way of drawing on personal experience, I assess whether or how change is actually occurring and can reveal the processes at work here, particularly those "hidden" processes (see Baigent, 2001). This starts with my own recognition that I am part of any masculine hegemony that I am studying. It also brings out some factors that I have either previously not considered or which my remaining allegiance to male firefighters may have caused me to previously ignore or suppress.

I begin by reflecting on my personal experience of joining the fire service in 1962 and living in a society that still largely accepted that women should be protected from dirty, physical and harrowing work in disasters. During the Second World War, when the London Fire Brigade was stretched to breaking point, male firefighters who had been called to military service were returned from the frontline to support their brothers instead of employing women to fight fires. This all but confirmed a tradition that still has credibility today and continues to deny rescue teams the full range of skills that a mix of genders would provide. This negative dynamic and the processes men use to reinforce it are examined below alongside the history of the employment of women in the UK fire service and examples from Australia and Sweden. I also discuss at some length a recent experience in Sweden that challenged me personally, before moving toward a final discussion about the remarkable resilience of gender regimes in the fire service.

My insider's view

When I joined the fire service in 1962, all firefighters were men, mostly with military backgrounds. Tradition had established a belief that it is not only men but "special" men who can provide the necessary mental strength, the physical body and the trust to form up in bonded and hierarchical teams to work safely and successfully in dangerous and messy disaster situations. This "tradition" though was manufactured by older men who, as I suggest in my PhD thesis (Baigent, 2001), were able to use homosocial practices (Lipman-Blumen, 1976) and bullying or harassment as a way of fitting people in or rejecting them (see also Baigent, 2004; Allaway, 2010). As an 18-year-old, I found it difficult to be accepted myself: The process of proof that you were tough enough was new to me, including the ability to stand up to the relentless joking used to police the masculinized behavioral norms, and there were many times that I almost gave up. However, once accepted, I too took on the role of gatekeeper as if it were the right and natural way to behave, until at 50 I retired and went to university. Then my first degree in sociology led to my questioning my own behavior, just as my PhD research later led me to unpack masculine cultures in the fire service.

In brief, my research was a qualitative study involving interviews, observation and profeminist autocritique (Hearn, 1994). This methodology has been very successful for me as a "gamekeeper turned poacher" because it allowed me to use my tacit knowledge to add to traditional analysis of data and to reveal secrets that men keep among themselves: a process rigorously tested by my own thesis (Baigent, 2001). As time passes and knowledge is shared about informal cultural arrangements (Yoder and Aniakudo, 1997; Ericson, 2011), the argument that the processes used to fit me in were an "innocent secret" are increasingly recognized instead as an unspoken arrangement kept within the fraternity by dramaturgical loyalty (Goffman, 1959, p. 212). This is the process that I now hope to unpack further.

Close up: landmarks of change in women's employment in the UK fire service

As argued earlier, there has been no great appetite among male firefighters to upset the gender ordering of their work, nor the dividend and superiority this gave them in the hierarchy of men (Pacholok, 2013). It should not be surprising, therefore, that 1960s equality legislation was ignored until the Equal Opportunities Commission forced the managers of the UK fire service to employ the first woman firefighter in 1982. One woman joined over 60,000 men. Now, over 30 years later, the percentage of full-time women firefighters remains at a token level of around 3% (Department of Communities and Local Government, 2014); thus leaving a situation where, because women unhinge popular beliefs about masculinity and the traditions that male firefighters adhere to, a lone woman's presence is significant. In similar terms, New York's Fire Service also accepted their first women in 1982 (LaTour, 2012). This dynamic is not unique, because in many countries the 3% ceiling is still the norm (NFPA, 2012). I will now introduce some of the interpersonal dynamics that help explain the persistence of this near total male domination.

"Women are not strong enough" and sexual harassment

At the forefront of men's reasoning were arguments that I have heard hundreds of times: "Women lacked the physical strength to do the job." The resultant lack of trust of women then underpinned the second argument "that women's presence would disrupt the strong bonds that existed among men." To an extent, and this is rarely acknowledged, these two arguments have a basis of truth.

In an organization that should have been able to see the difficulties with an existing physical entrance test, which almost assumed that men applying for the fire service would be strong enough, the rush to get women on board led to employing some women who did lack the physical strength needed to handle heavy equipment. Particularly, given that UK firefighters need four people to handle their ladders, if a women could not hold up her end it was feared that in an emergency a lack of strength could become an important safety factor. Eventually, this situation was addressed by making some equipment lighter, discarding other equipment and developing a job-specific physical test that recognized a new common standard. In hindsight, it should have been easy to identify just how many difficulties women who joined this all-male domain would experience. Neither is it beyond reason to suggest that women were set up to fail, either deliberately or carelessly in the "rush" to get women into the fire service. This led to stories—now folklore—of women firefighters as unreliable. This is a situation that still haunts the equality agenda.

Several high profile sexual harassment cases followed (Hearn and Parkin, 1995) and women joined and left in what became a constantly revolving door (Walby, 1990). As a consequence, harassment of women firefighters reached levels in excess of 60% (Baigent, 1996), and a fire service accustomed to solving problems was "caught in the headlights." At the same time, people from academia (Chetkovich, 1997; Yoder and Aniakudo, 1997), consultants (Corporate Communications Company, 1999) and the fire service itself (see Howell, 1994, 1996; Baigent, 1996; Webb, 1997; Archer, 1998) started to report on sexism in the fire service. Inevitably, equality was pushed up the agenda. For many managers, it became "*the* agenda" and in a fire service that was difficult for anyone to join, prioritizing the employment of more women added to men's belief that women were a problem.

The "waking up" phase: institutional responses

Toward the end of the twentieth century, in what might be called a "waking up" phase, the UK fire service responded with the appointment in 1997 of a new Chief Inspector of Fire Services, Graham Meldrum, who had an established positive equality profile. Shortly after appointment, he declared in a landmark report that the fire service was institutionally sexist (HMCIFS, 1999). Well-meaning, but sometimes questionable, measures, such as attempting to demilitarize and demasculinize by establishing new core values (LGA, 2004), had hardly any effect on firefighters' behavior—other than to increase backlash against women. The spending of £10 million to provide separate changing, washing and sleeping arrangements for women (ODPM, 2006) had the mixed effect of improving women's privacy, but at the same time caused a further male backlash as areas of male space were given over to women. An attempt to improve women firefighters' numbers through achieving a target of 15% female firefighters by 2009 (Straw, 2000) again caused backlash as male firefighters claimed that men were being passed over in favor of women.

The "waking up" phase: informal cultural responses

Despite the attention placed on institutional behaviors and the accompanying compulsory education programs, there remained a continual failure to convince all firefighters and managers to change their behavior. In some ways, blaming the institution for its sexism allowed male firefighters to shift the blame for their own difficult behaviors.

Previously thought to be solely responsible for institutional behavior, the formal culture of the fire service was interrogated through my own research (Baigent, 2001), revealing that, despite firefighters' claim that they join the fire service in order to serve, everything about firefighters' social relations pointed to something far deeper. This suggestion came from data that had first shown that the very real practical skills of firefighting were traditionally handed down to new firefighters by older firefighters. Second, I argued that, because this arrangement complemented and paralleled the aims of the organization, managers did not see the authority this gave older firefighters as a threat, but as a subculture within the formal culture.

It was these two points about the unofficial handing down of skills that my research examined in greater depth. The outcome was to draw attention to the probability that this cultural arrangement also handed down knowledge about social arrangements, particularly informal hierarchies. The positive of this process was that it bonded the team and gave a sense of belonging. However, on the negative side, this process was also closely aligned to exclusionary masculine agendas, specifically to a self-serving element for male firefighters' own purposes of identity and patriarchal dividend as an elite masculinity (see also Pacholok, 2013).

My reflexive view is that this informal cultural arrangement was an established mechanism in fire services long before women were considered eligible to join.

This process, including the informal testing of individual men to weed out those who did not fit in, was applied to me when I joined in 1962: it was accepted as a norm by both firefighters and managers at the time, and in many cases still is today. It was this behavior that my research recognizes as homosocial (Lipman-Blumen, 1976; Cockburn, 1991)—an analysis that came from an acknowledgement that many of these arrangements were inherently about men accepting male hierarchies, proving their masculinity and ensuring that others around them do likewise. These homosocial exclusionary practices included as legitimate the constant drip of harassment to drive out those who did not fit in, and to ensure newcomers accepted informal cultural arrangements and were "approved" before they were able to learn the physical and social skills of the masculinized vision of a fire service.

I also suggested that policing this arrangement included a considerable and ongoing self-appraisal, a three-part Foucauldian gaze turned inward which can and did "lead to firefighters believing their image and acting out at work how they subjectively judge they expect to be seen, by themselves, their peer group and the public" (Baigent, 2001, p. 101). There can be two purposes to this behavior. Firstly, it helps form a close- knit operational team that allows men to perform their dangerous work "safely," largely because they know they are being supported by like-minded firefighters whom they trust. Secondly, by being seen as a trusted comrade, firefighters actually come close in their own eyes to achieving the prize of being superior men in the hierarchy they establish through comparative strategies of the self: the three-part Foucauldian gaze in which they are their own judge and jury (Baigent, 2001; Pacholok, 2013).

In an all-male fire service used to such arrangements, little of this behavior was seen as problematic; that is, until the 1970s when principal managers tried to cut the size of the fire service. It was then, I argue, that principal managers broke the rules of the game and damaged their links to firefighters' subculture by joining with politicians who were attempting to cut the amount of firefighters and their appliances, which by default would reduce the service that firefighters could provide. Reducing firefighters' ability to serve also reduced their means of doing masculinity (Baigent 2001; see also Hall *et al.*, 2007). However, largely because so many middle managers were also against the cuts, there was no schism, and subcultural arrangements for handing on knowledge in a united fire service continued. But then promoting the employment of women in the 1990s became a key aim of principal managers and politicians, and this time the middle managers were forced to join in. The changes they tried to implement as a consequence which were to actively set out to employ women and to change the way the men worked in their all male environment thus put them in direct opposition to the key subcultural arrangements by which firefighters proved themselves as special men: the fact that the whole workforce and their environment was male.

Given that male firefighters' whole identity had been premised on the fact that they were male and masculine, their resistance was all but inevitable, especially since firefighters' informal cultural arrangements provided a clear base for resistance to women's employment. I therefore identified a schism that put the subculture out of the control of managers. This then led to a stand-alone and separate informal culture to protect one of the last jobs where men can set themselves apart from others in a male-positive and acclaimed way (Baigent, 2001, 2004).

Positives and negatives

My 2001 suggestion that there was a second culture in the fire service, one largely out of control of the managers, was something of a challenge to an organization where managers were accustomed to thinking they had all the answers. The Fire Service College, the entity responsible for training senior managers, was, and potentially still is, skeptical about these empirical findings on informal cultures. They rejected my research as experiential and partial, but it is my contention that the fire services' failure to address its second culture is a major reason standing in the way of achieving equality. A recent report (Lucas, 2015) provides clear evidence of how bad bullying and harassment currently is.

An international culture

During this and subsequent research, it emerged that firefighters' behavior was not necessarily specific to a particular shift or a fire station; each shift may have had its own particular quirks. However, the central notions can be universally recognized in all fire services around the globe. Older men claim and are given authority by younger men; and male firefighters' work teams act as a primary reference group in regard to how they do their work, who does it and how this affects the image of a firefighters as manly heroes (Baigent, 2001; Cooper, 1986). All of my subsequent research reinforces the view to suggest that, whether on a fire station in the USA, Australia or in the UK, you would face this same cultural requirement to fit in with whatever local version of this informal cultural definition prevailed, not because this was necessarily a requirement to be a good firefighter, but because this is how trust is established (Allaway, 2010; Desmond, 2007; Ericson, 2011).

Comparative interventions and responses: the power of men's informal cultures in firefighting

The fire services record on the employment of women is not an easy read for an equality activist. There have been a number of interventions aimed at teaching firefighters how to behave. I look to Sweden for an answer.

Sweden

An expectation throughout the world is that Sweden is at the top of the league in regards to women's rights. However, in 2009 there was some alarm in Sweden that their fire service had so few women firefighters. In an attempt to align the fire service to the rest of their society, I was asked to work with the Swedish Fire Service. Given what effectively was a "clean slate," I suggested that Sweden reject the dominant model in which women firefighters were dispersed throughout the service. I suggested instead that Sweden try to bring women together in centers of excellence where resources could be focused on cultural change. Sweden accepted this model (see MSB, 2009) and one service was chosen to implement what became known as a "fire station for all" (Baigent *et al.*, 2012).

A central feature of this work involved a team of five providing an educational package to help firefighters to accept change. My role was to provide a hard-hitting message of how I had recognized my own sexism, using examples directly from the fire service to talk with the Brandmen, as firefighters are called in Sweden. I further suggested that, direct and overt harassment apart, harassment in the fire service generally occurs through a constant "drip" of criticism that points out to women that they are not wanted. To emphasize cultural complicity in this, I included in the training the idea that, when this behavior is observed and not challenged, the observer is also part of the harassment (Baigent *et al.*, 2012; Baigent and Granqvist, 2013).

We had many Brandmen who "confessed" to difficult beliefs, particularly when we spoke of a recent example of harassment drawn from their fire service. However, the more we repeated the course, the harder our message became to deliver. Increasingly, there were some Brandmen who made the whole concept very personal, creating a "standoff" that unhinged and diverted the training program. This involved Brandmen turning themselves into victims by arguing that they were being stereotyped, even bullied by the trainer/educator. Ironically, but not unexpectedly, some women in these training sessions supported their male colleagues and the process may have alarmed managers into becoming wary of this training approach. This was not the first time I had experienced this harsh reaction. It may be a new tactic by firefighters to overcome equality training—an approach in total opposition to earlier acknowledgements by the fire service that it was sexist.

Australia

An example from Australia shows the power of male firefighters. Many years after research in Melbourne showed that women firefighters were being harassed (Lewis, 2004), a Gender Inclusion Action Plan was produced (MFB, 2010). The union directly challenged the notion that women were harassed in their fire service. More than this, they successfully took action to stop the implementation of the new policy, in part by using the voices of women firefighters who joined with their union to deny sexism existed (Schneiders, 2010). As I had been part of some of this work, I was drawn into the debate and sought to help salvage the program (Baigent, 2010); this was unsuccessful and the union's militancy was sufficient to stop the whole program.

A new phase

There is a trail of evidence that some women are now being accepted, and growing evidence that men may be adapting their behavior. Male as well as female critics are standing up in the new climate created by advocates to promote women's entry into firefighting, some arguing that examples of bad behavior do not apply to them. That this could happen makes clear that male firefighters are now hearing the message and change, albeit slowly, is occurring. But this may only be an accommodation, and if men are hiding their sexism, what does this mean for lasting change? It may be that men's beliefs about women are now better hidden, as when men recognize that it is right to stop being sexist when a woman is in the room—but when she steps out, they can still believe it is acceptable to talk about their sexual activity, laugh at sexist jokes and watch pornography (Baigent, 2008). What appears to be happening is that male firefighters have learned how to behave when they are among women, again demonstrating the power of their informal and separate workplace culture. In some respects, this is progress. Men's "surface acting"—knowing how to behave while among women (Hochschild, 1983)—may in the long run turn to "deep acting" that supports fundamental change. My own experience and the evidence I have reviewed from Sweden, Australia and the UK suggest that equality remains a complicated social arrangement that has yet to be fully encompassed by male firefighters.

There is, though, one change that may be occurring as an unintended consequence. In the UK, firefighters are increasingly resisting managers by taking strike action to stop cuts to their service and conditions. During these strikes, women have proved themselves loyal comrades. From attending picket lines and marches, I have observed a relaxation of tensions as women are clearly proving themselves a necessary part of the team. Could it be that, due to managers' increasing attempts in recent times to cut the fire service and change firefighters' conditions, the cutting of the fire service has become a more important issue than gender? This would be a real irony if firefighters unite in opposition under the umbrella of their union to oppose cuts that threaten their service. When women firefighters joined with men on the picket lines, they proved they were "one of the boys." Gender became of less concern to firefighters who stood together against what was seen as a greater enemy. These are early days, but women's inclusion alongside their male colleagues—and indeed the leading role they are often taking at such times—may bring a reduction in harassment and greater acceptance.

Conclusion

My very close association with the fire service continues though, time after time, I experience disparity between the talk and the practice. My original idea in this chapter was to reflect on the ways that men had resisted change in the fire service, hoping this might help provide a means to improve the gender balance of the teams that respond to disasters.

However, it cannot conveniently be forgotten that male firefighters challenge everyone who joins their team, and the requirement that they prove themselves through the exclusionary practice of homosociality (Baigent, 2001). I am now somewhat more skeptical that male firefighters' behavior is in fact hegemonic. I recognize that hegemony can be flexible, but the change underway in the fire service may not be a hegemonic adaption; instead, it may be an authentic if partial acceptance of women. Could it be that while firefighters' image may support hegemonic masculinity, it only does so superficially? Might some male firefighters genuinely suspend their exclusive attitudes when a woman who proves her capability joins their team? It may be that women who meet the grade are granted and accepting of honorary masculine status. But, if we accept that masculinity is a social construct (Collinson and Hearn, 1996; Connell, 2001), then maybe we can equally recognize that women, too, can be masculine (Lorber, 2000) though the celebration of difference has been a major platform for women's groups in the fire service.

From the examples of men's outrage above, I can identify the possibility that acceptance is happening in Sweden, though not from these examples in Australia. There is clearly room for further research. These examples from the picket line that show women may be more accepted because they can be trusted to fit in industrially. This raises some issues about gender being subordinate to class in a Marxist sense that space limitations do not permit me to examine here.

Masculinity as an expression of male power may be in the process of transformation in the fire services of the twenty-first century although the default expectation remains that firefighters are male. While men will accommodate women, it appears that comradeship during industrial action may be speeding this up. Yet, men in the fire service are unlikely to forget the powerful cultural traditions of their past, keeping these alive and well in their heads and in the informal culture that nourishes them. The losers here are not only women. The general public is also denied the wide-ranging skill base that a mix of genders provides when firefighters come to their aid at disasters.

References

Allaway, B. (2010) Exploration of culture and change in the Scottish Fire Service: The effect of masculine identifications. MA, University of Edinburgh.

Archer, D. (1998) *Report of results from an equal opportunities survey.* Unpublished report for Hereford and Worcester Fire Brigade.

Baigent, D. (1996) Who rings the bell? A gender study looking at the British Fire Service, its firefighters and equal opportunities. MA. Anglia Ruskin University.

Baigent, D. (2001) Gender relations, masculinities and the fire service: A qualitative study of firefighters' constructions of masculinity during firefighting and in their social relations of work. PhD. Anglia Ruskin University.

Baigent, D. (2004) Fitting-in: The conflation of firefighting, male domination, and harassment. In: Gruber, J. and Morgan, P. (eds.) *In the company of men: Re-discovering the links between sexual harassment and male domination.* Boston: North Eastern University Press, pp. 45–64.

Baigent, D. (2008) One decade on: Data on the harassment of women in the UK Fire and Rescue Service. Unpublished manuscript. Available from http://www.fitting-in.com/decade/harassment.doc (accessed December 16, 2015).

Baigent, D. (2010) Fighting fires or fighting male hegemony? *Online Opinion*, March 26, post. Available from http://www.onlineopinion.com.au/print.asp?article=10224 (accessed December 16, 2015).

Baigent, D., Brunzell, L. and Granqvist, L. (2012) "En brandstation för alla." [A fire station for all]. Available from http://www.rsyd.se/om-oss/vara-projekt/bfa (accessed December 16, 2015).

Baigent, D. and Granqvist, L. (2013) Stopping the drip. Paper delivered at the MSB and GRO Conference on Equality. Stockholm.

Carrigan, T., Connell, R. and Lee, J. (1985) Toward a new sociology of masculinity. *Theory & Society*, 4(5), pp. 551–604.

Chetkovich, C. (1997) *Real heat: Gender and race in the urban fire service*. New Brunswick, NJ: Rutgers University Press.

Cockburn, C. (1991) *In the way of women*. London: Macmillan.

Collinson, D. and Hearn, J. (1996) *"Men" at "work": Multiple masculinities/multiple workplaces*. In: Mac an Ghaill, M. (ed.) *Understanding masculinities*. Buckingham: Open University Press, pp. 61–76.

Connell, R. (2001) The social organisation of masculinity. In: Whitehead, S. and Barrett, F. (eds.) *The masculinities reader*. Cambridge: Polity, pp. 30–50.

Cooper, R. (1986) Millais's *The Rescue*: A painting of a "dreadful interruption of domestic peace." *Art History*, 9(4), pp. 471–486.

Corporate Communications Company (1999) *Report of research for West Midlands Fire Service concerning the 1999 recruitment strategy*.

Department of Communities and Local Government (2014) Fire and rescue: Operational statistics bulletin for England: 2013–14. Available from https://www.gov.uk/government/uploads/system/uploads/attachment_data/file/347567/Operational__Statistics_Bulletin_2013-14.pdf (accessed December 16, 2015).

Desmond, M. (2007) *On the fireline: Living and dying with wildland firefighters*. Chicago: University of Chicago Press.

Ericson, M. (2011) Up close: Masculinity, intimacy and community in firefighters' work teams. PhD. University of Gothenburg.

Goffman, E. (1959) *The presentation of self in everyday life*. New York: Doubleday Anchor Books.

Hall, A., Hockey, J. and Robinson, V. (2007) Occupational cultures and the embodiment of masculinity: Hairdressing, estate agency and firefighting. *Gender, Work and Organization*, 14(6), pp. 534–551.

Hearn, J. (1994) Research in men and masculinities: Some sociological issues and possibilities. *Australian and New Zealand Journal of Sociology*, 30(1), pp. 40–60.

Hearn, J. (2004) From hegemonic masculinity to the hegemony of men. *Feminist Theory*, 5(1), pp. 49–72.

Hearn, J. (2012) A multi-faceted power analysis of men's violence to known women: From hegemonic masculinity to the hegemony of men. *The Sociological Review*, 60(4), pp. 589–610.

Hearn, J. and Parkin, W. (1995) *"Sex" at "work": The power and paradox of organization sexuality*. London: Prentice Hall.

Her Majesty's Chief Inspector of Fire Services (HMCIFS) (1999) Equality and fairness in the fire service: A thematic Review by HM Fire Service Inspectorate. Home Office report. Available from http://www.nwcfbu.co.uk/equality/equality_thematic_review_1999.pdf (accessed December 16, 2015).

Hochschild, A. (1983) *The managed heart: Commercialisation of human feeling*. Berkeley: University of California Press.

Howell, M. (1994) Women firefighters: "The inequality gap." MBA. University of Hertfordshire.

Howell, M. (1996) Fire service culture, asset or burden? MA. Fire Service College.

LaTour, J. (2012) *Sisters in the brotherhoods: Working women organizing for equality in New York City*. New York: Palgrave Macmillan.

Lewis, S. (2004) Gender and firefighter training. Research paper 2, Institute for Social Research, Swinburne University of Technology, Melbourne. Available at http://www.mfbb.vic.gov.au/asset/PDF/GenderProjectPaper2.pdf (accessed May 15, 2015).

Lipman-Blumen, J. (1976) Homosocial theory of sex roles. *Signs*, 1(3), pp. 15–31.

Local Government Association (LGA) (2004) Core values in the Fire and Rescue Service. Available at http://www.lga.gov.uk/Documents/Agenda/fire/180205/item2.pdf (accessed May 15, 2015).

Lorber, J. (2000) Using gender to undo gender. *Feminist Theory*, 1(1), pp. 79–95.

Lucas, I. (2015) Independent cultural review of Essex County Fire and Rescue Service. ECFRS report. Available from http://www.essex-fire.gov.uk/_img/pics/pdf_1441197562.pdf (accessed December 16, 2015).

Metropolitan Fire and Emergency Services Board (MFB) (2010) Gender inclusion action plan. Available from http://www.fitting-in.com/australia/melbourne/article%20for%20the%20age/Gender_Inclusion_Action_Plan_4%5B1%5D.pdf (accessed December 16, 2015).

National Fire Protection Association (NFPA) (2012) Firefighting occupations by women and race. Available from http://www.nfpa.org/research/reports-and-statistics/the-fire-service/administration/firefighting-occupations-by-women-and-race (accessed December 16, 2015).

ODPM (2006) Supplementary memorandum by the Office of the Deputy Prime Minister (FRS 31(d)).

Pacholok, S. (2013) *Into the fire: Disaster and the remaking of gender*. Toronto: University of Toronto Press.

Schneiders, B. (2010) Fire brigade chief aims parting shot. *The Age*. March 23. Available from http://www.theage.com.au/victoria/fire-brigade-chief-aims-parting-shot-20100322-qrdv.html (accessed December 16, 2015).

Straw, J. (2000) *Equal opportunities in the fire service: Targets for the recruitment retention and career progression of women*. London: Home Office.

Swedish Civil Contingencies Agency (MSB) (2009) Programme of action for increased equality and diversity in municipal safety work: For the period 2009 to 2014. Report. Available from http://www.fitting-in.com/sweden/Swedish%20Programme%20of%20action.pdf (accessed December 16, 2015).

Walby, S. (1990) *Theorizing patriarchy*. Oxford: Basil Blackwell.

Webb, J. (1997) The politics of equal opportunity. *Gender, Work and Organization*, 4(3), pp. 159–169.

Yoder, J. and Aniakudo, P. (1997) Outsider within the firehouse: African American women firefighters. *Gender & Society*, 11(3), pp. 324–342.

16 Using a gendered lens to reduce disaster and climate risk in Southern Africa

The potential leadership of men's organizations

Kylah Genade

The extent of gender inequality in Southern Africa has undermined good governance, constricted development, compromised human welfare and increased social vulnerabilities contributing to disaster risk. It has particularly disempowered girls and women. As a result, the primary focus of gender equality movements has been directed toward women and girls (Barker and Ricardo, 2005; World Economic Forum, 2013). But, this focus overlooks the interdependence of men and women and the interrelationships that exist between and among men as a group. Prioritizing women has also deflected the call to address the specific needs of boys and men from a gender perspective. However, in Southern Africa many efforts are underway to change gender stereotypes and encourage gender equality. Male-led community-based organizations have used participatory dialogue and other inclusive practices to promote open discussions between men and women.

This chapter reflects on how men's social justice work may in future help mitigate the effects of disasters and climate-related risk by reducing gender inequalities. I begin by exploring the social constructions of manhood in Southern Africa and the building blocks of disaster risk reduction and climate change adaptation, then present regional examples of men's organizations working to explore gender relations in many domains. The case for men's solidarity work with women to help address climate and disaster risks using a gendered lens is developed.

The analysis draws on relevant literature that describes promising male-led initiatives and on original case material. Empirical support is provided by interview data from the Integrated Adolescent Girls in Community Based Disaster Risk Reduction in Southern Africa Project (IAG), led by the author in 2012 and 2013 in Lesotho, Malawi, Zambia and Zimbabwe (Genade and van Niekerk, 2014). Data were gathered from 14 focus groups held with IAG participants (both boys and girls); 120 interviews with the participants' parents; and structured interviews with 28 men and 36 women from the Ikageng township in North West Province, South Africa.

Constructs of manhood and masculinity in Southern Africa

In the Southern African context, men are the main beneficiaries of the uneven gender divide, wherein they most often derive the majority of power, access and resources (World Economic Forum, 2013). Efforts to challenge or reorient the roles ascribed to men and the ideals associated with manhood and masculinity in Southern Africa require a significant shift in thinking and action on individual, structural and institutional levels.

Boys become men through reward as well as sanction. Gender roles traditionally portray men as the holders of powerful positions in Black Southern African society (Redpath *et al.*, 2008). The positions are often accompanied with decision-making authority and superiority in communities and family units (Brown *et al.*, 2005). This power has been vested across generations through patriarchal laws that prioritize men and grant them leadership as fathers and husbands, and as elders and tribal chiefs (Morrell, 1998). Great respect is traditionally paid to African warriors who played heroic roles leading battles and defending kingdoms, and who owned herds of cattle and fathered numerous children from their multiple wives (Hunter, 2004; Cornwall and Lindisfarne, 2005). This contrasts starkly with views of Black women in Southern Africa that emphasize support and servitude to the family with strong domestic caregiving, food preparation, cleaning and childrearing responsibilities (Geisler, 1995; Shefer *et al.*, 2007).

Further, certain subcultures, such as those of the South African Xhosa people, regard the status of manhood as not achieved through physical development alone but through participation in customary initiation schools. Here, teenage boys become men by formally learning their roles and responsibilities (Peltzer and Kanta, 2009); following male circumcision, each is adorned in a blanket, receives a stick and is formally referred to as an *amakwala* ("new man"). The completion of this practice entitles the new man to marry, own property and be a respected speaker in public (Peltzer and Kanta, 2009, citing Meintjies, 1999). Conversely, failure brings social shunning, lack of public respect and inability to partake in ceremonies such as traditional marriages (Morrell, 1998).

The nature of risk and gender in Southern Africa

Disasters as unique phenomena are born from the combination of hazard exposure and vulnerability, while overwhelming the capacities of those affected (United Nations International Strategy for Disaster Reduction, 2004). Climate change refers to documented longitudinal changes in the state of the climate (United Nations International Strategy for Disaster Reduction, 2009). These changes bring about increased frequency and severity of extreme weather conditions (hydro-meteorological hazards) which can directly increase disaster risk (O'Brien *et al.*, 2006).

Southern Africa is a region encompassing great diversity in its geography, culture, language and economies. In 2012, the area was home to approximately 167 million people with approximately two-thirds of the population considered to be poor and living on less than US$1 a day (Southern African Development Community Secretariat, 2008). In addition to gender inequalities, hunger, HIV/AIDS, malnutrition and marginalization contribute to the high social vulnerability of the population in the face of climate change and disasters (Southern African

Development Community Secretariat, 2008). Disaster risk for Southern Africa is shaped primarily by exposure of this highly vulnerable population to a myriad of natural hazards (Mulugeta *et al.*, 2007), all compounded by the effects of climate changes which lead to threats of increasingly frequent extreme hydro-meteorological hazard events, extreme temperatures and weather patterns (Climate and Development Knowledge Network, 2012). Drought and drought-related hydro-meteorological hazards are the primary causes of economic loss and mortality in the Sub-Saharan Africa region (Dilley *et al.*, 2005).

As a result of the complex nature of disasters, impacts are not uniform across populations and often evident along gender lines (Enarson and Morrow, 1998a; Neumayer and Plümper, 2007). The risk posed from disasters has the potential to influence developmental goals and to accentuate factors contributing to vulnerability such as gender inequality. Social power is a defining factor in all societies and is distributed along gendered lines. Nowhere are the imbalances of power more evident than in the context of disasters, where differential impacts on men, women, boys and girls have been widely documented.

It is important to acknowledge that men and women have very different roles, responsibilities and power, and that these factors will assist in building or undermining their ability to address the risk they face (Wisner *et al.*, 2002). Gender and gender relations can influence people's ability to expect, plan, endure, cope with and recuperate from disasters based on their differing roles, responsibilities and access (Valdes, 2002; Rashid and Shafie, 2013). To be effective, efforts to reduce risk and address climate change adaption must therefore acknowledge the underpinnings of power and power distribution along gender lines as an underlying contributor to risk. Failure to adequately account for discrepancies in the effects of disasters on both men and women can further reinforce their respective vulnerabilities and detracts from efforts to address critical dynamics involved in shaping the escalation of risk (Bolin *et al.*, 1998).

Applying a gendered lens to risk reduction and climate change adaptation is a way of illuminating social processes associated with disaster and climate change adaptations, which can serve to reduce or to reinforce inequalities (Enarson and Morrow, 1998b). Gender analysis is one way to appreciate the gendered dimensions of risk in a given population. The critical first step is to take into account the unique context of Southern Africa, understanding men's and women's distinctive lived experiences, the underlying causes of the differential impacts they experience, and the coping strategies and adaptation they adopt. These factors bear strongly on gendered risk and may also support gender-aware climate adaptation strategies and disaster risk reduction plans.

Different lived experiences

The daily lives of Southern African men and women vary across gender lines, as do the experiences they have that are grounded in risk perception. When participants from peri-urban Lusaka (Zambia) were asked about the things they felt threatened by in their hazard-prone communities, boys cited "floods, poor toilets and poor structures (buildings), cholera, sickness, malaria, droughts, floods, and blocked drainages" while girls spoke of such threats as "bars, uncovered wells," "drug abuse, and the drunkards." Zambian boys similarly prioritized environmental and biological hazards while girls feared social threats more. This suggests that adaptation strategies and risk reduction programs should not settle for a standard approach but build on risk as perceived differently by boys and girls.

Different underlying causes

Neumayer and Plümper, in their international study of gender gaps in disaster-related fatalities, documented that "socially constructed vulnerabilities derive from the social roles men and women assume, voluntarily or involuntarily, as well as existing patterns of gender discrimination" (2007, p. 551). In Southern Africa, customary laws restrict land ownership to men, see children as men's property and limit women's financial independence. Value is assigned through characteristics such as physical strength, often associated with roles of leaders and warriors linked to manhood and masculinity. In Zimbabwe, for instance, ideals surrounding men and boys are linked to physical power; as boys explained, "boys can fight back." Sotho boys (Lesotho) also stated that "boys have strength to fight." Zambian boys clearly acknowledged differences in power between themselves and girls, wherein "boys are stronger, they can protect themselves" and women "are weak and unempowered."

The limited expectations for girls in society based on gender roles associated with domestic work and motherhood are used to justify the withdrawal of girls from school. In Zimbabwe, it was said, "some parents stop paying school fees for a girl child and she has to drop out of school, but they pay for boys." Perceptions of the limited value or worth of girls and women in Malawi society are similar: "When it comes to educating girls, people believe that when you educate girls, they just end up being prostitutes."

For their part, girls recognized in our interviews that boys possess power that is often used against girls. For instance, girls in Zimbabwe described boys as "perpetrators" and as "gangsters that threaten girls." These negative perceptions of men and boys often relate back to the realities experienced by girls victimized by male-initiated violence. Girl respondents from all four countries studied spoke of the male threat of "battery, beating women, rape, and violation." In Malawi, girls claimed "boys don't have the threats we face. They are free." These profound differences in power and vulnerability and the social distance they create can overshadow shared concerns, leaving men and boys unprotected and vulnerable in disaster contexts. These data present a recognized difference in the risk perception of men and boys compared to women and girls. However, this realization and subsequent consideration must be seen against the backdrop of previous research that acknowledges that men have a tendency toward more risk behavior as evident in works such as that of Harris, Jenkins and Glaser (2006).

Southern African men's organizations

The roots of colonialism and patriarchal culture in Southern Africa have fostered gender inequality. Yet, importantly, the recent rise in men's organizations has led to the opening of dialogue among men, and between men and women, on critical issues such as inequality, gender roles, sexual decision-making and domestic violence. Their work helps to focus the gender lens on the social processes which inadvertently reinforce disaster-relevant vulnerabilities along gender lines.

The 1995 Fourth United Nations World Conference on Women in Beijing served as a foundation for promoting equal rights but also recognized gender equity as a potential new global goal. Three years later, men in the Southern African region started to take the lead in their own movement to end violence against women (Sonke Gender Justice Network, 2011). Male-led marches in South Africa, the formalization of men's organizations in Zambia and Namibia, and men's initiatives at the community level have sought in particular to target the elimination of gender-based violence (Ericksen, 2007). There has been a strong call for male involvement in matters surrounding reproductive health and gender-based violence, and men-focused and male-led initiatives have been seen as the new vehicle for addressing gender inequality in the region (International Centre for Research on Women Asia Regional Office, 2007; Redpath *et al.*, 2008; Stern *et al.*, 2009).

This forward movement has encouraged the rise and advancement of additional Southern African based men's organizations with priorities in areas including health, HIV/AIDS and parenting (Mehta *et al.*, 2004; Sonke Gender Justice Network, 2011; Barker *et al.*, 2011; Hazangwi, 2014). The South African Men's Forum, for instance, has worked on issues of brutality against women and children, while Sonke Gender Justice supports "One Man Can" community-level branches to promote gender equality through sensitizing and reeducating men and boys (Mhangawi-Ruwende, 2014). The "5 in 6 Project" based in Cape Town also deals with understanding the roots of violence against women. Additionally, MenEngage actively involves men as agents of positive change regarding the abuse and victimization of women in communities in Zambia, Namibia, Malawi and South Africa (Stern *et al.*, 2009). PADRE/Enkundleni, Men's Forum on Gender, MWENGO, African Father's Initiative, Student and Youth Working on Reproductive Health Action Team (SAYWHAT), ZIYON and Ecumenical Support Services have all been working in Zimbabwe to target male violence against women/girls and reproductive health related issues (Sonke Gender Justice Network, 2013).

Many of these organizations have led interventions such as workshops, training, capacity building initiatives and participatory dialogues aimed at encouraging positive behavioral change (Stern *et al.,* 2009; Sonke Gender Justice Network, 2011; Hazangwi, 2014). Their activities are reflective of the need to understand the guiding social processes and norms which shape the lives of men and women in their communities. Through this participation, education and awareness, more men are coming to understand gender imbalances and the implications of these for community-wide health, education, ownership of resources, rights and responsibilities (White *et al.*, 2003).

These men's organizations illustrate the important part they can play in helping to understand the many different roles ascribed to men in Southern Africa and the implications of these for interpersonal relationships and community dynamics. How men benefit from and/or are negatively affected by power allocation supporting gender-based violence and their dominance in sexual decision-making is one significant entry point for larger dialogues around the distinctive social vulnerabilities and capacities of people in the face of hazards and disasters. Men's organizations directly draw attention to issues of sex and gender and through this work help to create a forum for dispelling myths, challenging beliefs and developing new ways of thinking about gender equality.

Entry points for reducing disaster and climate risk

While environmental justice work has not been a priority in the Southern African context historically, the gender and social justice work of these grassroots efforts bears closely on community-driven risk reduction campaigns, highlighting structural transformations related to gender that must be recognized and changed.

More specific outcomes may, in the long run, bear directly on disaster risk. Many men's organizations are engaged in activities that open dialogue between men and women, and host discussions regarding gender roles and culture and those promoting responsible parenting. The opportunity for creating shared understanding and changing attitudes can serve as a means for addressing the underlying factors contributing to the vulnerability of women in Southern Africa. This supports Wisner *et al.* (2002) in that a series of underlying factors or root causes increase vulnerability and hence disaster risk. Disparities in power, resource ownership and access are examples of how gender inequality can inhibit the capacity of women to withstand the impact of hazards.

The work of men's organizations can be used to challenge and break down the limiting structures and power ideals that shape risk. Specific actions directed toward dispelling stereotypes, creating new attitudes and encouraging positive behavior change toward gender equality, can be a starting point for minimizing these root causes. Gender roles which allocate the burden of domestic work to women have negative implications for women's differential risk exposure. The time spent doing unpaid labor and the social isolation caused by being tied to domestic caregiving for dependants (the elderly, children, disabled persons and the ill) ultimately contribute to their increased vulnerability. This is an entry point for the work of organizations such as Sonke Gender Justice Organization in South Africa, which has conducted community workshops on the concepts of manhood and the role of gender inequality as a contributing factor to high HIV rates (Urdang, 2008; Stern *et al.*, 2009). Their work reveals that attitudinal change is possible. The project director reported that after the third day of a workshop in Hoedspruit (South Africa), an elderly male participant revealed: "Yesterday, after I got home, I called my sons, I called my wife, and I explained to them what we are doing in this workshop" (Urdang, 2008, p. 2). He told his children that things had to change in their home. No longer could their mother arrive home tired from a day of work and be expected to cook, clean, wash the dishes and clear up all on her own. It was simply unfair.

Many women's initiatives in Southern Africa focus on capacity building and livelihood development, environmental conservation, agricultural diversification, disaster preparedness and resilience building. These workshops have encouraged the notion that women and men can go beyond traditional stereotypes and work toward creating more equitable home situations with shared responsibilities. Men's social justice organizing may also enable men to better understand the dimensions and demands of family life, and to have more significant involvement in domestic roles, allowing women more time to engage in external activities. This may include more employment outside of the home, expanding and diversifying the household's income and thus reducing disaster vulnerability.

Following from these changes, women's self-esteem and confidence, as well as their additional time freed from exclusive domestic responsibilities, allow their more active participation in formal or informal disaster preparedness associations. More social interaction and networking with peers, relatives, neighbors and friends is another positive outcome, allowing for improved access to hazard warnings and information about preparedness practices. Research by Perry and Lindell (1986) reaffirms the value of social networks in influencing the engagement of protective measures which can reduce vulnerability.

Activities aimed at aiding dialogue among men and women can further assist in building trust and respect between the groups. This outcome of men's organizing affords the opportunity for understanding differences in risk perception, a gender pattern particularly relevant to disaster risk reduction as the differences often cut across gender lines (Cutter *et al.*, 1992; Harris *et al.*, 2006). Similarly, creating more dialogue and potentially stronger trust relations between men and women can promote greater equality in household decision-making and potentially shift underlying norms of male dominance and authority. Women and girls clearly benefit; moreover, such shifts give women leverage for negotiating with men around such difficult issues as men's tendency to overlook warnings or take greater risks in hazardous conditions (Harris *et al.*, 2006). If women are given the credibility and respect necessary for their views to be considered, men might take warnings, threats and the need for responsive action more seriously, thus creating safer environments for boys and men in disaster contexts.

Mutual respect as a foundation for strong healthy relationships also has positive implications for improving men's health, for instance as women's greater influence might encourage men to seek needed medical interventions to reduce their vulnerability derived from physical illness (Department of Health and Ageing, 2010). Cultural stereotypes of men and masculinity in Southern Africa foster beliefs that men must control and not express emotion (Barker *et al.*, 2011). Thus, disasters can create psychological distress among men over loss of lives, property and livelihoods. Often times this distress manifests itself in violence, excessive use of alcohol and such high risk behaviors as unprotected sex (Sonke Gender Justice Network, 2013). Men's organizations can encourage men to express emotions within safe group settings and emphasize that stronger partner relationships are a positive way of sharing life's challenges and minimizing the gendered sense of responsibility among men for the negative impacts of hazards and disasters.

Although these connections are currently more aspirational than realized in practice, the work of men's organizations such as Sonke Gender Justice clearly illustrates the potential entry points of men's gender and social justice organizing for broader disaster risk reduction movements.

Conclusions

This chapter has sought to present the successes of men's social justice organizations in Southern Africa in a way that suggests how these might well contribute to addressing the environmental and social challenges of reducing disaster and climate risk. The narratives of boys and girls at high risk in the region clearly indicated the negative power of traditional gender roles in the face of hazards and disasters. Social inequality along gender lines serves to exacerbate disaster risk. The achievement of gender equality is a prerequisite for achieving the full representation necessary for effective disaster and climate-related risk reduction activities. As such, there is great potential for the expansion of men's organizations, and greater linkages with the work of women's groups in the Southern Africa region to promote disaster risk reduction and climate change adaptation.

References

Barker, G., Contreras, J.M., Heilman, B., Singh, A.K., Verma, R.K. and Nascimento, M. (2011) Evolving men: Initial results from the international men and gender equality survey (IMAGES). Washington, DC: International Centre for Research on Women. Available from http://www.icrw.org/sites/default/files/publications/Evolving-Men-Initial-Results-from-the-International-Men-and-Gender-Equality-Survey-IMAGES-1.pdf (accessed December 20, 2015).

Barker, G. and Ricardo, C. (2005) Young men and the construction of masculinity in Sub-Saharan Africa: Implications for HIV/AIDS, conflict and violence. Washington, DC: Social Development Department, The World Bank. Available from www.eldis.org/vfile/upload/1/document/0708/DOC21154.pdf (accessed December 20, 2015).

Bolin, R., Jackson, M. and Crist, A. (1998) Gender inequality, vulnerability and disaster: Issues in theory and research. In: Enarson, E. and Morrow, B.H. (eds.) *The gendered terrain of disaster: Through women's eyes.* Westport, CT: Praeger, pp. 27–44.

Briceño, S. (2002). Gender mainstreaming in disaster reduction. Commission on the Status of Women presentation, March 6. New York. Available from http://worldbank.mrooms.net/file.php/349/references/Salvano_Mainstream_Gender_ISDR_CSW_6_March_02.pdf (accessed December 21, 2015).

Brown, J., Sorrell, J. and Raffaelli, M. (2005) An exploratory study of constructions of masculinity, sexuality and HIV/AIDS in Namibia, Southern Africa. *Culture, Health and Sexuality,* 7(6), pp. 585–598.

Climate and Development Knowledge Network (2012) Managing climate extremes and disasters in Africa: Lessons from the IPCC SREX Report. London: Climate and Development Knowledge Network. Available from http://www.ifrc.org/docs/IDRL/-%20To%20add/ManagingClimateExtremesAfrica.pdf (accessed December 20, 2015).

Cornwall, A. and Lindisfarne, N. (eds.) (2005) *Dislocating masculinity: Comparative eth-nographies*. London: Routledge.

Cutter, S., Tiefenbacher, J. and Solecki, W.D. (1992) En-gendered fears: Femininity and technological risk perception. *Organization and Environment*, 6(1), pp. 5–22.

Department of Health and Ageing (2010) National male health policy—Building on the strengths of Australian Males. Barton: Commonwealth of Australia. Available from http://www.health.gov.au/internet/publications/publishing.nsf/Content/building-strengths-males-foreword (accessed December 20, 2015).

Dilley, M., Chen, R., Deichmann, U., Lerner-Lam, A., Arnold, M., Agwe, J., Buys, P., Kjevstad, O., Lyon, B. and Yetman, G. (2005) *Natural disaster hotspots: A global risk analysis*. Washington, DC: World Bank.

Enarson, E. and Morrow, B.H. (eds.) (1998a) *The gendered terrain of disaster: Through women's eyes*. Westport, CT: Praeger Publications.

Enarson, E. and Morrow, B.H. (1998b) Why gender? Why women?: An introduction to women and disaster. In: Enarson, E. and Morrow, B.H. (eds.) *The gendered terrain of disasters: Through women's eyes*. Westport, CT: Praeger Publications, pp. 1–8.

Ericksen, A. (ed.) (2007). Defying the odds: Lessons learnt from Men for Gender Equal-ity Now. African Women's Development and Communication Network (FEMNET). Available from http://femnet.co/2015/11/09/defying-the-odds-lessons-learnt-from-men-for-gender-equality-now/ (accessed February 15, 2016).

Geisler, G. (1995) Troubled sisterhood: Women and politics in Southern Africa: Case stud-ies from Zambia, Zimbabwe and Botswana. *African Affairs*, 94(377), pp. 545–578.

Genade, K. and van Niekerk, D. (2014) A new protocol in disaster risk reduction policy and praxis for the Southern Africa region. Gender-Age Socio-Behavioural Interventions and the GIRRL Programme Model. In: Perera, S., Henrikson, H.J., Revez, A. and Shklovski, I. (eds.) *Proceedings of Second Residential ANDROID Doctoral School in Disaster Resil-ience*. Newcastle upon Tyne: University of Northumbria, pp. 65–74. Available from http://nrl.northumbria.ac.uk/21198/1/RDS_2014_Proceedings_-_v2.2.pdf (accessed December 20, 2015).

Harris, C., Jenkins, M. and Glaser, D. (2006) Gender differences in risk assessment: Why do women take fewer risks than men? *Judgment and Decision Making*, 1(1), pp. 48–63.

Hazangwi, K. (2014) Zimbabwe—MenEngage boys and men for gender equality. Avail-able from http://menengage.org/regions/africa/zimbabwe/ (accessed April 2015).

Hunter, M. (2004) Masculinities, multiple-sexual-partner, and AIDS: The making and unmaking of Isoka in KwaZulu-Natal. *Transformation*, 54, pp. 123–153.

International Centre for Research on Women Asia Regional Office (2007) Engag-ing men and boys to achieve gender equality: How can we build on what we have learned? Washington: International Centre for Research on Women. Available from http://www.icrw.org/files/publications/Engaging-Men-and-Boys-to-Achieve-Gender-Equality-How-Can-We-Build-on-What-We-Have-Learned.pdf (accessed December 20, 2015).

Mehta, M., Peacock, D. and Bernal, L. (2004) Men as partners: Lessons learned from engaging men in clinics and communities. In: Ruxton, S. (ed.) *Gender equality and men: Lesson from practice*. Oxford: Oxfam, pp. 89–100.

Mhangawi-Ruwende, B. (2014) International: A call to men to fight gender violence. Available from http://www.genderlinks.org.za/article/international-a-call-to-men-to-fight-gender-violence-2014-11-27 (accessed December 21, 2015).

Morrell, R. (1998) Of boys and men: Masculinity and gender in African studies. *Journal of Southern African Studies*, 24(4), pp. 605–630.

Mulugeta, G., Ayonghe, S., Daby, D., Dube, O.P., Gudyanga, F., Lucio, F. and Durrheim, R. (2007). Natural and human-induced hazards and disasters in Sub-Saharan Africa. Pretoria: International Council for Science. Available from http://www.icsu.org/icsu-africa/publications/reports-and-reviews/icsu-roa-science-plan-on-hazards-disasters/Doc%20SP03.1_ICSU%20ROA%20Science%20Plan%20-%20Hazards%20and%20Disasters.pdf (accessed December 11, 2015).

Neumayer, E. and Plümper, T. (2007) The gendered nature of natural disasters: The impact of catastrophic events on the gender gap in life expectancy, 1981–2002. *Annals of the Association of American Geographers*, 97(3), pp. 551–566.

O'Brien, G., O'Keefe, P., Rose, J. and Wisner, B. (2006) Climate change and disaster management. *Disasters*, 30(1), pp. 64–80.

Peltzer, K. and Kanta, X. (2009) Medical circumcision and manhood initiation rituals in the Eastern Cape, South Africa: A post intervention evaluation. *Culture, Health and Sexuality*, 11(1), pp. 83–97.

Perry, R.W. and Lindell, M.K. (1986) *Twentieth century volcanicity at Mt. St. Helens: The routinization of life near an active volcano*. Tempe: Arizona State University.

Rashid, A.K.M.M. and Shafie, H.(2013) Gender and social exclusion analysis in disaster risk management. In: Shaw, R., Mallick, F. and Islam, A. (eds.) *Disaster risk reduction approached in Bangladesh*. Tokyo: Springer, pp. 343–363.

Redpath, J., Morrell, R., Jewkes, R. and Peacock, D. (2008) Masculinities and public policy in South Africa: Changing masculinities and working towards gender equality. Cape Town, Sonke: Sonke Gender Justice Network. Available from http://menengage.org/wp-content/uploads/2014/06/Masculinties_and_Public_Policy_in_South_Africa_FINAL_250509.pdf (accessed December 20, 2015).

Shefer, T., Ratele, K., Strebe, A., Shabalala, N. and Buikema, R. (eds.) (2007) *From boys to men: Social constructions of masculinity in contemporary society*. Lansdowne, Western Cape: UCT Press.

Sonke Gender Justice Network (2011) The MenEngage Africa Network regional organization capacity survey. Johannesburg: Sonke Gender Justice Network. Available from http://www.genderjustice.org.za/publication/the-menengage-africa-network-regional-organisational-capacity-survey/ (accessed December 21, 2015).

Sonke Gender Justice Network (2013) Good practice brief on male involvement in GBV prevention and response in conflict, post-conflict and humanitarian crisis settings in Sub-Saharan Africa. Johannesburg: United Nations Population Fund, Eastern and Southern Africa Regional Office. Available from http://www.genderjustice.org.za/publication/good-practice-brief-on-male-involvement-in-gbv-prevention-and-response/ (accessed December 11, 2015).

Southern African Development Community Secretariat (2008) Comprehensive care and support for orphans, vulnerable children and youth (OVCY) in the Southern African Development Community. Gaborone: Southern African Development Community. Available from http://www.sadc.int/files/2113/5293/3505/SADC_Strategic_Framework_and_Programme_of_Action_2008-2015.pdf (accessed December 20, 2015).

Stern, O., Peacock, D. and Alexander, H. (eds.) (2009) Working with men and boys: Emerging strategies from across Africa to address gender-based violence and HIV/AIDS. Johannesburg: Sonke Gender Justice Network. Available from http://menengage

.org/wp-content/uploads/2014/01/MenEngage-Africa-Symposium-Case-Studies-1.pdf (accessed December 20, 2015).

United Nations International Strategy for Disaster Reduction (2004) *Living with risk.* Geneva: United Nations. Available from http://www.unisdr.org/files/657_lwr1.pdf (accessed December 20, 2015).

United Nations International Strategy for Disaster Reduction (2009) UNISDR terminology on disaster risk reduction. Available from http://www.unisdr.org/we/inform/terminology (accessed December 11, 2015).

Urdang, S. (2008) Enlisting men for women's equality. South African initiatives against sexual violence gender inequities. *Africa Renewal*, 22(1), pp. 6–8. Available from http://www. genderjustice.org.za/download/enlisting-men-for-womens-equality/?wpdmdl=2011 (accessed December 20, 2015).

White, V., Greene, M. and Murphy, E. (2003) Men and reproductive health programs: Influencing gender norms. Available from http://pdf.usaid.gov/pdf_docs/Pnacu969.pdf (accessed December 11, 2015).

Wisner, B., Blaikie, P., Cannon, T. and Davis, I. (2002) *At risk: Natural hazards, people's vulnerability and disasters*. London: Routledge.

World Economic Forum (2013) The global gender gap report 2013. Geneva: World Economic Forum. Available from http://www3.weforum.org/docs/WEF_GenderGap_Report_2013.pdf (accessed December 20, 2015).

17 Training Pacific male managers for gender equality in disaster response and management

Stephen Fisher

Any work that attempts to gender disaster management and response must confront the issue of engaging professional men. Gender training for predominantly male disaster practitioners has bloomed since the mid-1980s (Seed, 1999; Bradshaw, 2015) and continues to be seen as the solution to the sexism or gender blindness of organizational practices and programs. As Seed (1999, p. 311) notes, it is premised on the idea that "if only planners and development practitioners could be fully 'gender-aware,' then projects could be planned which would benefit women as well as men." Gender awareness training is often promoted as the preferred change strategy; thus, this chapter critically examines training interventions with men to improve their support for gender equality in disaster response and management.

Writing on gender, development and disaster, Bradshaw notes that "attempts to engender disasters appear to be a number of decades behind processes to engender development" (2013, p. 58). Unfortunately, some writers who do write in this vein about men and disaster draw on a men's liberationist gender perspective criticized as undermining women's rights (Messner, 1998). Mishra, for example, asserts that "boys and men are ... constrained by current gender stereotypes regarding male roles and masculinity which makes it difficult for men to be different" (2009, p. 31). Such a perspective in effect erases gender power relations (Connell, 1987). Instead, it blames male socialization for a pervasive insecurity felt by men which purportedly results in greater likelihood of aggression and self-hatred, particularly during times of social instability. The concepts of gender role modeling and socialization theory are problematic, not only for lacking recognition of power relations, but also for the tendency to reduce the complexity of gender relations to two different, yet otherwise equal, categories of people (Connell, 1987, 2000; Pease, 2002).

Too often, the inclusion of gender into aid agencies' planning and programming is condensed to identifying and describing characteristics, and thus needs, of "two fixed categorical binaries, women and men, where the complexity of gender is tamed by a pre-existing normative taxonomy" (Pacholok, 2013, p. 28). For example, a Red Cross Fact Sheet on Gender and Disaster Management refers to the way "gender shapes the extent to which men, women, boys and girls are vulnerable to and affected by emergencies and disasters" and further that "understanding that men and women face different obstacles can help the development of more effective programs and ensure that needs are really met"

(Red Cross, 2010). This implies a neutral force that impacts equally on gendered individuals who lack agency.

This chapter explores the tensions associated with providing gender training for disaster managers by highlighting the shortcomings; although this sort of work is potentially a useful strategy in need of further development. I critically interrogate my own training experience in this domain and closely examine eight contemporary programs aimed at developing gender sensitivity among disaster practitioners. These training materials were prepared by individuals from a range of international government and nongovernment organizations (GOI and UNDP (India), 2008; Pincha, 2008; Community Based Disaster and Circle, 2009; Enarson, 2009; Ciampi *et al.*, 2011; Rahman and Agrawal, 2012; IIRR and Cordaid, 2013; WHO Pakistan and GOP, 2013).

The limitations of gender training

Representing the classical Gender and Development ideology, Kabeer's (1994) critical analysis of gender training remains relevant today (Porter, 2012; Bradshaw, 2013; Schwarz, 2014). Kabeer notes that, while there are often similarities in gender training approaches and content, they can vary in terms of three main types of emphasis and/or objectives. Some programs include a clear personal component aiming to challenge entrenched attitudes, while others are explicitly political. Kabeer's third type of goal for training involves equipping professionals with the skills and knowledge required to undertake gender analysis, and therefore able to include gender as a cross-cutting issue in all aspects of the program management cycle. Often considered technocratic (March *et al.*, 2003), the emphasis on "gender tools" is commonly included in these approaches (among others, see Pincha, 2008; Enarson, 2009; Rahman and Agrawal, 2012; WHO Pakistan and GOP, 2013). One manual, for example (Community Based Disaster and Circle, 2009), refers to gender-based analysis as an analytical tool, while another (Ciampi *et al.*, 2011) relies heavily on Anderson and Woodrow's Capacity and Vulnerability Analysis tool to assess the "different needs, capacities, and vulnerabilities of women and men" (1989, p. 82).

Carolyn Moser is well known among gender and development practitioners, particularly for the development of the popular Moser Gender Planning Framework (Moser, 1993) and therefore has been influential in the way gender training has developed. Mukhopadhyay and Wong (2007) argue Moser's approach to training contains contradictory positions. On the one hand, she acknowledges that training is only one strategy advocates can use and that achieving gender equality requires political intervention. On the other hand, Mukhopadhyay and Wong (2007, p. 16) assert, Moser is optimistic in making ambitious claims that short training sessions will allow participants not only to easily grasp complex concepts but also to develop the ability to undertake complex gender analysis. They further explain that this confidence is evidenced in Moser's concern that gender training does not promote tension by raising personal or political issues, but stays focused on professional, technical skills.

Gender is introduced in common ways across all the sample gender and disaster training programs under review here: First, it is used to explain the distinction between sex and gender; and, secondly, it is framed as an ahistorical and universal phenomenon. For example, the Training Curriculum on Women Leadership in Disaster Risk Reduction (Rahman and Agrawal, 2012, p. 12) explains that "Society exhibits a division of labor between men and women. It is a social construct; and it shapes perception about men and women. It results in allotting one set of roles to women and another set to men." Third, gender is commonly introduced as a way to identify different needs or issues for the two gender categories. For example, the Community Based Disaster and Circle (2009, p. 91) refers to identifying "activity profiles of men and women before, during and after the disaster to analyse changes in gender-based division of labor and needs of men and women." There is scope to develop more complex analyses of gender relations both within training curricula and within the training context.

Rather than the issue being an overly optimistic expectation of gender training, or the lack of a multifaceted set of interventions, Standing (2004) explains that the problem has been that gender advocates do not understand how to engage effectively with policy machinery. The assumption is often made that gender training can be transformative by equipping participants with clear and simple analytical concepts, such as "gender roles and responsibilities," and "practical and strategic" gender needs (Molyneux, 1985). While these concepts are underpinned by feminist theorizing, "they can equally be problematic in reducing the complex to the banal and seeming to promise the riches of political change without the long work of politics" (Standing, 2004, p. 82). Similar concerns have been voiced by gender trainers themselves (Porter and Smyth, 1998). Mukhopadhyay and Wong (2007, p.13) point out that "training has been understood as one of a number of key gender strategies but insufficient by itself." Particularly in the case of working with men, my experience has been that it is effective when part of a range of organizational change strategies. While recognizing that there are strategic limitations to gender training, there remains the problem of inappropriate or overly simplified training content.

Applying gender training frameworks in disaster management

While some of the manuals examined explicitly mention the need to change or challenge the traditional gendered division of labor, this is marginal to the central theme of the training which appears to be the more technical planning aim of identifying gender-disaggregated statistics and characteristics for development planning, without discussing the relational aspects of gender, power and ideology. There appear to be two key aims in teaching disaster response managers about the gendered division of labor. The first is to challenge the taken-for-granted reality that genitalia determine roles and responsibilities by replacing this "natural attitude" dichotomy with an understanding of the dynamic and socially constructed nature of gender (Wharton, 2011). The second aim is to uncover the discrimination and inequality that arise from this humanly constructed arrangement. These two conceptual challenges are crucially important where men are concerned.

At this point, it appears that the eight comparable gender and disaster training guides and my own Fiji workshop referenced in this discussion have a number of shortcomings in common, albeit in varying degrees:

1   short time frames for learning complex concepts;
2   a tendency, in spite of inclusion of discussion-based activities, to be more pedagogically didactic, based on a "banking model" of education, than participatory with emphasis on mutually constructive learning;
3   a technocratic focus on application of tools that primarily result in gender descriptions rather than analysis of gender relations; and
4   a lack of a specific focus on men's privilege and power.

I have written elsewhere about these significant limitations in terms of training men for gender equality (Fisher, 2014). However, here I want to distinguish the Fiji training that I was involved in from the other approaches surveyed in one small, yet significant way. As the participants were comprised solely of men from the Pacific, not only did their gender mean that they were more likely to be resistant (Connell, 2003); it was also likely that they held to specific cultural norms and beliefs that reinforce the naturalness of the current gender order. This does not imply that resistance to gender equality based on cultural norms is peculiar to the Pacific context. Indeed, Hewitt makes the point that the "case for addressing cultural concerns ... is seriously undermined if one assumes that they just apply to ... underdeveloped, traditional, ethnic or backward societies" (2012, p. 90). At the risk of the charge of neocolonialism, the last part of this chapter presents one way the Fiji workshop sought to specifically deal with men's resistance and culturally based sexism through particular workshop processes and content.

Gender training with Pacific male disaster managers: a special challenge of culture and religion

I was engaged in 2011 to provide a one-day training in Suva, Fiji, to a group of 15 male Red Cross disaster managers from a range of Pacific countries, including Fiji, Solomon Islands, Cook Islands and Tuvalu. The stated aims of the course were: 1) to improve understanding of all staff of the meaning of gender integration and gender equality; 2) to develop deeper understanding of the way gender can be considered at all stages of disaster management; 3) to understand how cultural norms and beliefs can impact on gender in disaster management; and 4) to achieve a greater sense of how gender dimensions can be incorporated into each person's daily work.

While the training was well received and a range of engaging interactions were facilitated, I was left feeling ambivalent about the value of such interventions or "capacity building," a term criticized as neoliberal (Phillips and Ilcan, 2004). While the comparable training reviewed did address cultural issues (see below), the Fiji training with male practitioners had more of a direct focus on critically interrogating culture and religion, and sought to destabilize ideological supports for gender inequality. My focus therefore is on Kabeer's (1994) training goals of encouraging men to take on the personal and political dimensions of gender relations, not just their technical aspects.

Commonly, as found in the comparable programs, the conceptual distinction between biological sex and gender is "taught" to participants early in a training before moving fairly swiftly on to other substantive areas. However, I argue that men in particular hold very strongly to cultural and religious discourses that prevent women's rights (Ertürk, 2007). So only by subjecting these beliefs to intense and critical scrutiny will they be loosened. Such work is complex and time-consuming but also foundational, if the goal sought is organizational actors who will move toward becoming political gender equality advocates (e.g, see Casey, 2010; Stoltenberg, 2013).

The link between hegemonic masculinity and culture and religion needs to be recognized as important in effective disaster risk management. According to Hewitt, in spite of the lack of attention, culture is "integral to understanding risk and responses to danger or loss" (2012, p. 86). Cannon (2014, p.12) explains that the "interaction between culture and risk relates to many aspects of human and institutional behavior, including religious and related beliefs." In spite of the significant influence of religion, Schipper *et al.* (2014) observe that "Most international organizations and high income country participants (donor partners) in initiatives to reduce environmental risk (including climate change) are intentionally agnostic in their approach" (pp. 54–55).

However, as Connell suggests (1987, 2003), in addition to other explanations for men's resistance to gender equality, including men's material benefits and services, their fear of feminization and the resentment of low-status men regarding the focus on more privileged women, many men hold an ideological defense of male supremacy on the grounds of religion and cultural tradition. Walby (1990) similarly identifies ideological cultural supports as one of the six key aspects of patriarchy, a schema which also includes paid work, household division of labor, sexuality, violence and the state.

In disaster contexts, cultural norms are directly related to women's greater vulnerability to a range of hazards. Further, it is common that women may "accept their position and perceive it as being legitimate because it is regarded as cultural rather than exploitative" (Cannon, 2014, p. 21). This is particularly significant in the Pacific context for three reasons. First, there is a tendency for donor countries to demonstrate cultural sensitivity by focusing on the heterogeneity of traditional Pacific cultures (Anderson, 2012, p. 3) rather than on the commonality of gender inequality. For example, women leaders are rare in religious organizations across the Pacific (Swain, 1995). Secondly, it is useful to recognize that culture or customs are "a symbolic construction, a contemporary human product rather than a passively inherited legacy" (Linnekin, 1992, p. 249) and therefore the participants are agents in its challenge or reproduction. This can be seen in a meeting of a Pacific Women's network which developed strategies to challenge culture and tradition (Pacific Feminist SRHR Coalition, 2013).Thirdly, Pacific writer Tengan notes that the weight of tradition and custom is commonly employed by men in the Pacific to challenge (neo)colonialism, asserting instead a nationalism that tends to be "structured patriarchally, configuring the woman as the embodiment of tradition and mother of the nation which needs to be protected by militarized masculine men" (2002, p. 242).

This strong/weak gendered dichotomy is strongly rooted in both culture and religious discourse. It must be challenged as women are more at risk than men and have less access to aid and rehabilitation due to inequality in "decision-making and women's weak bargaining power within the household" (Mehta, 2007, p. ix). However, the only direct critique of culture, among the training manuals reviewed here, is offered in an Indian program which urges (without providing specific training content) that "a gender-aware facilitator would be alert and identify those cultural practices that create gender discrimination and violate rights of women" (Pincha, 2008, p. 81). The comparable gender and disaster risk management training programs surveyed tend to refer to culture in abstract terms, as an item to be "mapped" (Community Based Disaster and Circle, 2009) or suggest the need to adapt tools to be "sensitive" to particular cultural contexts (IIRR, Cordaid, 2013, p. 1). Similarly, a mapping approach is taken in a report relevant to the Pacific context which presents culture as a factor influencing gendered roles and responsibilities thus in need of recognition for planning purposes (UNDP Pacific Centre and AusAID, 2009). A Philippines-based program does explore the issue of culture further by acknowledging the reluctance of some tribal groups to promote the role of women. Demonstrating the way men in particular discursively employ the defense of "traditional culture" against neocolonialism to resist gender equality (Narayan, 1997), the trainers take note of male resistance (Community Based Disaster and Circle, 2009, pp. 156–157):

> They cautioned that this should not contravene their culture—a core issue in the struggle of the indigenous peoples. … a tribal chieftain … said that these activities for women should not teach wives how to slap their husbands, nor should they attempt to equalize the status of women.

The training methodology: developing criticality and responding to resistance

While the workshop brief for disaster risk managers in the Pacific required that I develop the men's gender analysis skills and ability to apply appropriate assessment "tools," it was agreed from the outset that a critical discussion of culture and religion would be valuable as these are essential components of effective disaster risk management institutions (Gopalakrishnan and Okada, 2007, p. 357).

Many of the gender and disaster risk management training curricula surveyed did not appear to have a clearly articulated pedagogical framework. While also not explicitly documented, the Fiji workshop employed Brookfield's (2011, p. 55) critical thinking pedagogy. Brookfield describes five elements essential in assisting students to develop a critical approach to social issues:

1   small group discussion where peers challenge each other;
2   teacher modeling the process of testing assumptions;
3   grounding the process in concrete experiences through case studies and scenarios;
4   being confronted with unexpected ideas or examples that invoke disorienting dilemmas as per Mezirow (2009); and
5   thinking critically, first about general issues at a distance, and gradually moving to critiques that are directly personally confronting.

Each of these elements was used in an attempt to challenge the taken-for-granted assumptions about the inviolability of culture and religion. This took the form of a range of questions designed to provoke discussion, disagreement and challenge. Specifically, the session on culture consisted of posing eight questions that the men were invited to discuss. In exploring each question, the men were asked to consider their own country's culture and history, and encouraged to identify specific examples to support their position. The intention was to draw the men toward certain conclusions, although disagreement was encouraged. The questions and intended conclusions were as follows:

1   Are cultures and traditions static? No; culture is constantly changing due to both internal and external influences.
2   Should culture be protected? Yes; certain rich meaning systems and practices should be maintained. However, those that infringe women's rights should be transformed. The intention is to break the culture-versus-gender equality dichotomy.
3   Is there agreement within a country about culture? No; every society experiences a range of contested views about what culture is, what should be preserved and what should change. There are dominant or ruling views and there are subordinate positions.
4   Is there a culture protection double standard? Yes; women are more likely to be considered the preservers or vessels of culture and often norms are applied more strictly to men than to women.
5   Are development initiatives ever culturally neutral? No; every time a new program or intervention is introduced it will have an impact on culture.
6   Should development initiatives ever try to be culturally neutral? No; this is impossible. Often so-called culturally neutral initiatives result in reinforcing or exacerbating gender inequalities. This is particularly relevant in the disaster response and readiness sector that can tend to value hegemonic masculine traits and approaches.
7   Isn't cultural sensitivity important? Yes; sensitivity is important. However, respect for cultures "is not merely an uncritical acceptance when culture, tradition or religion are invoked" (Schalkwyk, 2000, p. 3).
8   And what about men? At this stage, this question usually has already been discussed and men's use of culture as an excuse for maintaining inequality has been drawn out. I therefore use it as an opportunity to present examples

of men supporting gender equality without the perceived risk of cultural dis-integration. For instance, I present a case study of a Cambodian chief who has reportedly maintained his cultural status and respect while making significant changes in his personal life. Much rich discussion ensued upon hearing the chief's word: "I think I'm very successful at it," he grins, cuddling his two grandsons. "I feel we [men] can do everything, share everything—except giving birth and breastfeeding" (Hunter, 2009).

Similarly, the session on religion titled "Biblical Arguments for Gender Equality" provided provocative questions to enable debate and discussion. These included:

- Why do preachers ignore the majority of verses that are in favor of equality?
- If conflicting messages exist, why choose those supporting inequality?
- Why do many men selectively quote an isolated few verses that (seem to) promote inequality?
- Is it possible that we are misunderstanding the proper meaning of these few?
- Why are some passages seen as mandatory, while many are not adhered to as they are considered inappropriate or outdated?

The aim in unsettling the certainty of cultural and religious gender doctrines was directly related to positive outcomes for women in disaster preparedness and response. For example, it is essential that women's freedom of movement and participation in decision-making is supported. As Enarson and Chakrabarti note, the "full and equal participation of women and men is needed to mitigate haz-ards, reduce social vulnerabilities and rebuild more sustainable, just and disaster resilient communities" (2009, p. 42). While the general anecdotal feedback from the participants and Red Cross was positive, there was no formal evaluation; nor was there any opportunity to learn how the workshop impacted on participants' later practice. This chapter concludes with some tentative remarks about theories implicit in the short Red Cross training workshop with Pacific men.

Implications for developing gender training for men in disaster risk management

Working with men to address the gender dimensions of disaster risk management is very important. Not only are men more likely to hold key decision-making positions within government and NGO hierarchies, but also focusing on men serves to counter the common "feminization of responsibility" (Bradshaw, 2015) that so often emerges when institutions attempt to engender their work. When, in addition to gender power, male disaster managers have institutional power, albeit limited, for instance managers within a global corporate aid agency such as the Red Cross, critical approaches are even more necessary. Rao and Kelleher (2005, p. 66) make the point that organizational culture is significant as it can be a power-ful ally in making work on gender equality valued, or it can act as a key barrier. In a critique relevant to the highly masculinized work culture of disaster management,

Collinson and Hearn note that "men in organizations often seem preoccupied with the creation and maintenance of various masculine identities and with the expression of gendered power and status in the workplace" (1994, p. 3).

Developing conventional critical thinking skills, such as the ability to question assumptions, seek compelling evidence for conclusions and assessing the credibility of sources (Ennis, 1991), is a necessary pedagogical goal but insufficient for men who, by their privileged social location, are not inclined to acknowledge or undermine their gendered power. If the aim of training is to move as quickly as possible through to a recognition of the desirability of cultural change, then a sole focus on the development of critical thinking skills in relation to gender is unlikely to be sufficient. Burbules and Berk contrast the goals of critical thinking with critical pedagogy which sees "specific belief claims, not primarily as propositions to be assessed for their truth content, but as parts of systems of belief and action that have aggregate effects within the power structures of society" (1999, p. 47). The main difference, therefore, is the aim to challenge social injustice by transforming oppressive social relations and institutions.

According to Freire (1973), the task of critical pedagogy is to enable the oppressed group to gain insight into the nature and circumstances of their oppression. Importantly, in the case of male practitioners in disaster management, the focus is on the privileged rather than the oppressed. I have written elsewhere (Fisher, 2014) about the inherent tensions with applying Freirian theory to training men, not least of which is the tendency to "tread softly" when attempting to change men's gender views for fear of creating alienation or defensiveness.

Disaster risk management remains an overwhelmingly male-dominated area and one that is culturally masculinized as well (Fordham and Ketteridge, 1998; Tyler and Fairbrother, 2013). There remains an urgent need to develop a coherent and effective critical feminist pedagogy on gender equality for men working in both development and disaster risk reduction spheres. Only then will it be possible to apply realistic evaluation methods to deeper levels of intangible institutional change required (Rao and Kelleher, 2005, p. 62). This is even more pressing in disaster work because the stakes are high, with government and NGO agencies granting gender little concern and women being disproportionately harmed.

References

Anderson, C.L. (2012) Analysis of integrating disaster risk reduction and climate change adaptation in the US Pacific Islands and Freely Associated States. Available from http://www.pacificrisa.org/wp-content/uploads/2013/02/Anderson-Analysis-of-Integrating-Disaster-Risk-Reduction-and-Climate-Change-Adaptation.pdf (accessed August 29, 2015).

Anderson, M.B. and Woodrow, P.J. (1989) *Rising from the ashes: Development strategies in times of disaster.* Boulder, CO: Westview Press.

Bradshaw, S. (2013) *Gender, development and disasters.* Cheltenham: Edward Elgar Publishing.

Bradshaw, S. (2015) Engendering development and disasters. *Disasters*, 39(S1), pp. 54–75.

Brookfield, S. (2011) *Teaching for critical thinking: Tools and techniques to help students question their assumptions.* Chichester: Wiley.

Burbules, N.C. and Berk, R. (1999) Critical thinking and critical pedagogy: Relations, differences, and limits. In: Popkewitz, T.S. (ed.) *Critical theories in education: Changing terrains of knowledge and politics*. New York: Routledge, pp. 45–65.

Cannon, T. (2014) The links between culture and risk. In: Cannon, T. and Schipper, L. (eds.) *World disasters report: Focus on culture and risk*. Geneva: International Federation of Red Cross and Red Crescent Societies, pp. 12–35.

Casey, E. (2010) Strategies for engaging men as anti-violence allies: Implications for ally movements. *Advances in Social Work*, 11(2), pp. 267–282.

Ciampi, M.C., Gell, F., Lasap, L. and Turvill, E. (2011) Gender and disaster risk reduction: A training pack. Oxford: Oxfam.

Collinson, D. and Hearn, J. (1994) Naming men as men: Implications for work, organization and management. *Gender, Work & Organization*, 1(1), pp. 2–22.

Community Based Disaster Circle (2009) Integrating gender into community based disaster risk management. Philippines: Center for Disaster Preparedness. Available from http://www.preventionweb.net/files/14452_genderincbdrm1.pdf (accessed December 15, 2015).

Connell, R.W. (1987) *Gender and power*. Cambridge: Polity Press.

Connell, R.W. (2000) *The men and the boys*. St. Leonards, NSW: Allen & Unwin.

Connell, R.W. (2003) The role of men and boys in achieving gender equality. In: Consultant's paper for "The role of men and boys in achieving gender equality," expert group meeting, organized by DAW in collaboration with ILO and UNAIDS, pp. 21–24. Available from http://www.un.org/womenwatch/daw/egm/men-boys2003/aide-memoire. html (accessed December 15, 2015).

Enarson, E. (2009) Gender mainstreaming in emergency management: A training module for emergency planners. Toronto: Women and Health Care Reform. Available from https://www.gdnonline.org/resources/GEM_MainFINAL.pdf (accessed December 15, 2015).

Enarson, E. and Chakrabarti, P.G.D. (eds.) (2009) *Women, gender and disaster: Global issues and initiatives*. New Delhi: Sage.

Ennis, R. (1991) Critical thinking: A streamlined conception. *Teaching Philosophy*, 14(1), pp. 5–24.

Ertürk, Y. (2007) Intersections between culture and violence against women: Report of the Special Rapporteur on violence against women, its causes and consequences. Geneva: United Nations. Available from http://www.refworld.org/pdfid/461e2c602. pdf (accessed December 15, 2015).

Fisher, S. (2014) Involving men to end violence against women: A critical approach. PhD. Deakin University, Melbourne.

Fordham, M. and Ketteridge, A. (1998) "Men must work and women must weep": Examining gender stereotypes in disasters. In: Enarson, E. and Morrow, B.H. (eds.) *The gendered terrain of disaster: Through women's eyes*. Westport, CT: Praeger, pp. 81–94.

Freire, P. (1973) *Pedagogy of the oppressed*. London: Penguin.

GOI, UNDP (India) (2008) *Training of trainers manual on gender mainstreaming in disaster risk management*. New Delhi: Government of India and United Nations Development Program.

Gopalakrishnan, C. and Okada, N. (2007) Designing new institutions for implementing integrated disaster risk management: Key elements and future directions. *Disasters*, 31(4), pp. 353–372.

Hewitt, K. (2012) Culture, hazard and disaster. In: Wisner, B., Gaillard, J.C. and Kelman, I. (eds.) *Handbook of hazards and disaster risk reduction*. London: Routledge, pp. 85–96.

Hunter, K.D. (2009) By cooking and cleaning, Cambodia's Mr. Mom sweeps away poverty. Available from http://www.ilo.org/global/about-the-ilo/newsroom/features/WCMS_112996/lang--en/index.htm (accessed May 6, 2014).

IIRR, Cordaid (2013) Building resilient communities. A training manual on community managed disaster risk reduction. Philippines: International Institute of Rural Reconstruction and Cordaid. Available from https://www.cordaid.org/media/publications/Booklet_1_CMDRR_Training_Design_and_Implementation.pdf (accessed December 15, 2015).

Kabeer, N. (1994) *Reversed realities: Gender hierarchies in development thought*. London: Verso.

Linnekin, J. (1992) On the theory and politics of cultural construction in the Pacific. *Oceania*, 62(4), pp. 249–263.

March, C., Smyth, I. and Mukhopadhyay, M. (2003) *A guide to gender-analysis frameworks*. Oxford: Oxfam.

Mehta, M. (2007) *Gender matters: Lessons for disaster risk reduction in South Asia*. Kathmandu: International Centre for Integrated Mountain Development.

Messner, M.A. (1998) The limits of "the male sex role": An analysis of the men's liberation and men's rights movements' discourse. *Gender & Society*, 12(3), pp. 255–276.

Mezirow, J. (2009) *Transformative learning in practice: Insights from community, workplace, and higher education*. San Francisco: Jossey-Bass.

Mishra, P. (2009) Let's share the stage: Inclusion of men in gender risk reduction. In: Enarson, E. and Chakrabarti, P.G.D. (eds.) *Women, gender and disaster: Global issues and initiative*. Delhi: Sage, pp. 29–39.

Molyneux, M. (1985) Mobilization without emancipation? Women's interests, the state, and revolution in Nicaragua. *Feminist Studies*, 11(2), pp. 227–254.

Moser, C. (1993) *Gender planning and development: Theory, practice and training*. New York: Routledge.

Mukhopadhyay, M. and Wong, F. (eds.) (2007). *Revisiting gender training: The making and remaking of gender knowledge*. Oxford: Oxfam Publishing. Available from http://www.kit.nl/gender/wp-content/uploads/publications/1031_Gender-revisiting-web2.pdf (accessed February 16, 2016).

Narayan, U. (1997) *Dislocating cultures: Identities, traditions, and third-world feminism*. New York: Routledge.

Pacholok, S. (2013) *Into the fire: Disaster and the remaking of gender*. Toronto: University of Toronto Press.

Pacific Feminist SRHR Coalition (2013) Report of a strategy meeting of feminists advancing sexual and reproductive rights in the Pacific. Nadi, Fiji. Available from http://www.pacificwomen.org/wp-content/uploads/PACIFIC-SRHR-REPORT-FINAL-PUBLIC.pdf (accessed December 15, 2015).

Pease, B. (2002) *Men and gender relations*. Croydon, VIC: Tertiary Press.

Phillips, L. and Ilcan, S. (2004) Capacity-building: The neoliberal governance of development. *Canadian Journal of Development Studies*, 25(3), pp. 393–409.

Pincha, C. (2008) Gender sensitive disaster management: A toolkit for practitioners. Mumbai: Earthworm. Available from http://www.eldis.org/vfile/upload/1/document/0812/Gnder%20sensitive%20disaster%20management%20Toolkit.pdf (accessed December 15, 2015).

Porter, F. (2012) Negotiating gender equality in development organizations: The role of agency in the institutionalization of new norms and practices. *Progress in Development Studies*, 12(4), pp. 301–314.

Porter, F. and Smyth, I. (1998) Gender training for development practitioners: Only a partial solution. *Gender & Development*, 6(2), pp. 59–64.

Rahman, K.S. and Agrawal, M.K. (2012) Training curriculum on women leadership in disaster risk reduction. Bangladesh: National Alliance for Risk Reduction and Response Initiatives.

Rao, A. and Kelleher, D. (2005) Is there life after gender mainstreaming? *Gender & Development*, 13(2), pp. 57–69.

Red Cross (2010) Gender and disaster management. Available from http://www.redcross. org.au/files/2011_Gender_and_disaster_management_fact_sheet.pdf (accessed May 5, 2015).

Schalkwyk, J. (2000) Culture, gender and development co-operation. Available from http:// www.oecd.org/social/gender-development/1896320.pdf (accessed May 12, 2015).

Schipper, L., Merli, C. and Nunn, P. (2014) How religion and beliefs influence perceptions of and attitudes towards risk. In: Cannon, T. and Schipper, L. (eds.) *World disasters report: Focus on culture and risk*. Geneva: International Federation of Red Cross and Red Crescent Societies, pp. 36–63.

Schwarz, S. (2014) Critical perspectives on gender mainstreaming in disaster contexts. In: Zaumseil, M., Schwarz, S., von Vacano, M., Sullivan, G.B. and Prawitasari-Hadiyono, J.E. (eds.) *Cultural psychology of coping with disasters*. New York: Springer, pp. 323–342.

Seed, J. (1999) A history of gender training in Oxfam. In: Porter, F., Smyth, I. and Sweetman, C. (eds.) *Gender works: Oxfam experience in policy and practice*. Oxford: Oxfam, pp. 311–317.

Standing, H. (2004) Gender, myth and fable: The perils of mainstreaming in sector bureaucracies. *IDS Bulletin*, 35(4), pp. 82–88.

Stoltenberg, J. (2013) John Stoltenberg on manhood, male supremacy and men as feminist allies. *Feminist Current*. Available from http://feministcurrent.com/7977/john-stoltenberg-on-manhood-male-supremacy-and-men-as-feminist-allies/ (accessed December 8, 2013).

Swain, T. (1995) *The religions of Oceania*. London: Routledge.

Tengan, T.K. (2002) (En)gendering colonialism: Masculinities in Hawai'i and Aotearoa. *Cultural Values*, 6(3), pp. 239–256.

Tyler, M. and Fairbrother, P. (2013) Bushfires are "men's business": The importance of gender and rural hegemonic masculinity. *Journal of Rural Studies*, 30, pp. 110–119.

UNDP Pacific Centre, AusAID (2009) Stories from the Pacific: The gendered dimensions of disaster risk management and adaptation to climate change. Suva, Fiji: UNDP Pacific Centre. Available from http://www.preventionweb.net/files/9527_UNDPPCClimateChange1.pdf (accessed December 15, 2015).

Walby, S. (1990) *Theorizing patriarchy*. Oxford: Basil Blackwell.

Wharton, A. (2011) *The sociology of gender: An introduction to theory and research*. 2nd edn. Hoboken: John Wiley and Sons.

WHO Pakistan and GOP (2013) A training tool kit for community health workers on community based disaster risk management. Islamabad: World Health Organization. Available from http://applications.emro.who.int/dsaf/libcat/WHO_CBDRM_Trainers_ Guide_Book_EN.pdf (accessed December 15, 2015).

18 Integrating men and masculinities in Caribbean disaster risk management

Leith Dunn

This chapter argues the case for increased integration of men and masculinities in climate change and disaster risk management policies, programs and strategies. It exposes high risks that men at the margins of Caribbean society face when there is a natural hazard, because of socially constructed norms about masculinity. Conceptual, theoretical and practical issues are examined to support the need for this paradigm shift in disaster risk management in the Caribbean. The discussion first provides the context for understanding climate-related challenges in Caribbean Small Island Developing States. In the following section relating specifically to Jamaica, I highlight disaster-related vulnerabilities facing men at the margins of society: poor men, men with disabilities and gay men (Men who have Sex with Men, or MSM). Alternative theoretical frameworks on men and masculinities are presented with reference to disaster and climate risk, making the case for intersectional analysis. The chapter concludes with analysis of progress and gaps in gender mainstreaming in Caribbean disaster management agencies, and concrete recommendations for better integrating men and masculinities in national disaster policies, programs and strategies.

Caribbean Small Island Developing States: climate and disaster vulnerabilities

Arguably, integrating gender in climate and disaster risk management policies and programs in Caribbean Small Island Developing States (SIDS) will protect human rights and promote "smart" socioeconomic and sustainable development. However, this requires understanding of the vulnerabilities related to the region's complex, postcolonial political economy and cultural characteristics. Consideration must be given to the historical legacy of enslavement and globalization which are part of its postcolonial experience and which shape gender relations.

Caribbean countries face many climate-related challenges. They are among the group of 51 SIDS countries in the Atlantic, Indian and Pacific Oceans that share common characteristics and development challenges associated with their small populations, high population densities, open economies and limited resources. Many are remote from large land masses and markets and are vulnerable to hurricanes, droughts, tsunamis and volcanic eruptions. Many SIDS also have low economic resilience and are susceptible to global economic shocks. In Caribbean SIDS, the economic cost of climate change and disasters is high. In 2004, for example, several Caribbean countries experienced severe weather events which negatively impacted their economies. Damage from the 2004 hurricane season cost the Bahamas, Grenada, Jamaica and the Dominican Republic an estimated US$ 2.2 billion (UNDP, 2009). Grenada's damage from Hurricane Ivan included destruction of 90% of its hotel rooms, 80% of indigenous trees and 90% of its housing stock. Estimated economic losses were 38% of Gross Domestic Product (GDP).

Capacity to achieve long-term development goals declines, as limited financial resources are frequently diverted for post-disaster reconstruction. Caribbean Community (CARICOM) countries like Jamaica have reduced capacity to adapt to climate change because of Structural Adjustment Programmes (SAPs), the policy prescriptions imposed on indebted countries to reduce high levels of debt and stimulate long-term economic development. Despite heavy criticism of this model and its impact on Gross Domestic Product, SAPs continue to be prescribed by the International Monetary Fund and World Bank. Jamaica provides an example. Extremely high debt repayments (constituting over 50% of GDP) severely limit government's ability to fund climate change adaptation and disaster risk reduction programs, or provide financial allocations sufficient to meet basic health, education and other social services.

Poverty rates rose to 19.9% in 2012 from an already high rate of 17.69% in 2010. Poverty was 53.6% in male-headed households compared to 46.4% in female-headed households (Planning Institute of Jamaica, 2012). These data also show that single males were more likely than single women to be poor. Poverty and climate-related disasters and weather uncertainties have increased vulnerability at the household level and damaged physical infrastructure, further negatively impacting businesses and livelihoods. Together, these factors undermine national capacity to mitigate risks, recover from disasters and adapt to climate change.

The case of Jamaican men

Gender analysis of existing data on national vulnerability to climate and disaster risks can highlight the specific vulnerability of some groups of men, supporting the call for gender mainstreaming, an important but underdeveloped area when examining gender and disasters in Caribbean SIDS.

Poor men

Men and women in the poorest quintile of the economy are most vulnerable to the effects of natural disasters. The gender profile of this group includes: single unemployed men in low-income inner-city communities; single unemployed female household heads; poor men and women in rural areas; and men and women with disabilities, who have significantly higher levels of unemployment. These statistics show intersecting vulnerabilities and justify the need to integrate men and masculinities in disaster risk management. This approach would also result in an increased focus on men and women with disabilities.

Men with disabilities

Men and women with disabilities are often invisible but are vulnerable to disasters. Jamaica's Population and Housing Census (Anderson, 2009) reports 72,595 persons with disabilities: 36,088 males and 36,507 females. Their socioeconomic status and poverty levels are worse due to discrimination and less access to education and employment. Though data are scarce, gender and other intersecting vulnerabilities are likely to increase the challenges for both sexes in fulfilling their gender roles and responsibilities as family breadwinners and caregivers. Jamaica's Disability Act (Government of Jamaica, 2014) includes protection from discrimination but, as noted earlier, structural adjustment policies reduce the government's ability to fulfill these commitments or provide protective disaster risk management for men and women with disabilities.

Gay males

Men who have Sex with Men (MSM) in Jamaica are also vulnerable to disasters due to preexisting vulnerabilities that arise from heteronormative cultural norms and stereotypes. A recent national survey confirms that negative attitudes toward homosexuals persist across all sectors of society, while there have been some positive changes toward tolerance (Boxill *et al.*, 2012). These attitudes present barriers to MSM accessing their basic rights before, during and after a natural hazard event. Despite global commitments that recognize sexual rights as human rights, cultural attitudes pose specific challenges to national disaster risk management agencies and professionals responsible for implementing relevant policies and programs. For example, hurricanes and floods sometimes require displaced persons to spend time in temporary shelters. Gay men who seek refuge in these shelters may face discrimination and gender-based violence. The 2014 report of the Jamaica Association of Lesbians, Gays, Bisexuals and Transgender Persons noted that 80 incidents of discrimination, threats, physical attacks, displacement and sexual violence had been reported against LGBT persons in that year (J-FLAG, 2014, p. 14). Reports of male cross-dressers being assaulted by mobs are not uncommon.

Crime statistics confirm very low levels of reporting of gender-based violence of men against men; the rate of those reporting sexual violence against MSM is likely to be even lower than the reporting rates for female victims. Men raped by other men are unlikely to report these crimes to the police for fear of being stigmatized and ridiculed. Dominant hegemonic forms of masculinity imply that "real men" would not be victims of sexual attacks by men, hence the erroneous assumption is that any man who is sexually assaulted by another man is gay. Dunn and Sutherland (2009) analyzed data on victims and perpetrators of gender violence from police station diaries and through interviews with police officers, noting that reports of men being raped by men were low. This is not surprising as the legal framework is weak. The Offences against the Person Act only recognizes rape as males penetrating a female vagina. Advocacy to broaden the definition of rape to include a more gender-neutral definition has been resisted in Parliament and rejected by significant sections of the Jamaican society.

Gay men's preexisting risk of exposure to gender-based violence (GBV) is further illustrated by public response to recent media reports concerning a group known as "Gully Queens," who are gay men who live on the banks of a gully and in storm drains in the central commercial district of New Kingston. Many are homeless because of stigma and discrimination in their families, communities and workplaces. The Gully Queens are physically vulnerable as they live in storm drains and on the banks of the gully where heavy rain and extreme weather events such as a hurricane or flood increase their risk of being washed away. They are socially vulnerable as well, as they may avoid moving to a temporary shelter out of fear of being attacked or ostracized by other persons sharing the space. Efforts by government authorities to find them alternative housing have been hampered not only by stigma and citizens' unwillingness to accommodate the men in residential areas but also by the boisterous and criminal behavior of some members of the group. An interview with an insider indicated that homeless gay men come together like a family and protect each other.

Gay men who are poor and homeless are also especially vulnerable to HIV because stigma and discrimination against MSM encourage bisexuality, multiple sexual partnering and transactional sex (the exchange of money, goods and services for sex). These trends, and the limited access of MSM to treatment and services for HIV and other sexually transmitted infections, in turn increase HIV/STI risks in the wider population. Drug abuse among MSM can also contribute to high-risk sexual behavior. Unprotected sexual intercourse between MSM and exposure to sexual violence prior to, during and after disasters is a major development problem for Caribbean SIDS as the region has the second highest rate of HIV infection per capita globally, after countries in Sub-Saharan Africa. The National HIV program shows Jamaica's overall HIV prevalence rate is 1.7 %; however, among MSM the rate is 32%. The country's economic crisis and unemployment further increase the risk of transactional sex among MSM as a survival strategy.

Redefining masculinities: global and regional theoretical perspectives

The previous discussion on the vulnerability of men who are poor, have a disability or are gay demonstrates the need to redefine masculinities and broaden relevant theoretical frameworks to reflect a wider range of masculine identities. This redefinition can support gender mainstreaming, and help ensure that climate and disaster-related vulnerabilities are addressed in the high-risk male populations discussed above. New ways of thinking about men and masculinities may help explain and potentially reduce some of the hazards men face in disasters. For example, after Hurricane Mitch hit Honduras in 1998, there were reports of "heroic" actions by men in their protector role which put them in danger (BRIDGE, 2008; Bradshaw, 2004). The important link between gender and development patterns, and new hazards and disaster are highlighted by Bradshaw (2013, 2015). Men's risks and vulnerabilities in disasters and gender stereotypes need to be changed (Mishra, 2009). The approach to disaster management requires increased focus on issues affecting men and boys, including the gendered psychological impacts of losing their family, neighbors, assets, livelihoods, income and social power. Jonkman and Kelman's (2005) analysis of the causes and circumstances of flood-related deaths also justifies the need for increased focus on men and masculinities in disasters. Results from their analysis of 247 deaths in 13 floods in Europe and the USA showed that men accounted for 70% of flood-related fatalities. Drowning was the main cause, relating back to men's decision to drive or walk through floodwaters; taking unnecessary risks; and working in an emergency or support service role. Other causes related to men's exposure to physical trauma, heart attacks, exposure to fire, carbon monoxide poisoning and electrocution. Jonkman and Kelman's gender-sensitive methodology could usefully be applied in future to analyze gender mainstreaming policies and programs of the Caribbean Disaster Emergency Management Agency (CDEMA) and its member agencies.

Definitions of men and masculinities are being revised both globally and in the Caribbean in recent decades, including by Clatterbaugh (1990, pp. 9–12), who identifies six dominant theoretical perspectives on masculinity in the global literature. The Conservative Perspective (1) argues that it is natural for males to be providers and protectors; while the Profeminist view (2) rejects this claim and instead argues that masculinity is created through male privilege. This is oppressive to women and also harms men. Radical feminists in this vein argue that men use violence against women to oppress them and that masculinity exists in the context of the social and political order of patriarchy. Liberal profeminists argue that masculinity uses a system of rewards, punishments, ideals and social stereotypes to set limitations on men.

In contrast, the Men's Rights Perspective (3) argues that the traditional social role of men damages men, with emphasis on the need for new laws to protect men against injustice, especially relative to divorce, child custody and domestic violence. The Spiritual Perspective (4) holds that masculinity is derived from unconscious patterns, traditional beliefs, rituals stories and myths. Men, they argue, need to reach in themselves to rediscover themselves and grow. The Socialist Perspective (5), in turn, suggests that masculinity is grounded in economic and social class structures, primarily influenced by occupation and who controls labor and the products of labor. Their aim is to promote workers' control over their own labor and end ownership of the productive forces by private sector interests.

Finally, the Group Perspective (6), echoing lines of current intersectional analysis in feminist theory, provides a more inclusive and diverse framework and explanations for new forms of masculinity. This perspective acknowledges the role of racism and homophobia in the development of heterosexual masculinity. Black men, for example, acknowledge how their own masculinity is shaped by slavery, racism and poverty and White men's privilege; they seek to reduce the ways dominant White masculinity shapes anti-Black racism, just as gay men seek to reduce homophobia.

Intersectional analysis, a framework used extensively by feminists to analyze the multiple identities and forms of oppression experienced by women, highlights the importance of how social differences in power and social experiences can shape development. A comprehensive overview of the concept, its definitions, core principles and practical case studies of its value to policy-making, research on men's health and gender and climate change is available (Hankivsky, 2014). Country case studies confirm the value of using intersectionality to analyze power structures, processes and other inequalities, as well as understanding vulnerabilities and experiences of climate change. Theoretically, the Group Perspective's accommodations of inclusiveness and diversity provide, in my view, an effective framework for examining Caribbean men, masculinities and disasters. This would accommodate intersecting factors such as gender, age, race/ethnicity, class, sexuality and disability. Intersectionality, I further argue, can also help mobilize men to use their knowledge assets to mitigate disaster-related risks. MSM and other men at the margins of society can provide strategic knowledge and insights to mitigate risks associated with disasters.

Caribbean institutions mainstreaming gender in climate change and disaster risk management

Several Caribbean institutions are now working to implement global commitments to gender mainstreaming in climate change and disaster risk management. The Barbados-based Caribbean Disaster Emergency Management Agency (CDEMA) is the main regional organization guiding the process to implement a Caribbean Action Plan for Gender Mainstreaming in Disaster Risk Reduction. They support national disaster agencies to integrate gender in relevant policies, programs and strategies. In 2015, CDEMA implemented a project to further strengthen regional and national capacity to integrate gender in disaster management policies, programs and strategies. This included capacity to collect and analyze data, disaggregated by sex and other socioeconomic factors in order to highlight vulnerabilities of specific groups of men and women. The Belize-based Caribbean Community Climate Change Centre (CCCCC), a related organization, coordinates the region's response to climate change and promotes gender mainstreaming in their programs. Against this background, it is especially important that the policies, programs and strategies of disaster risk management agencies such as CDEMA and CCCCC, as well as such national equivalents as Jamaica's Office of Disaster Preparedness and Emergency Management (ODPEM), develop the capacity to analyze the intersecting realities of men in disaster risk management policies and programs.

In academia, the University of the West Indies (UWI), a regional university spanning 18 countries, plays a major role in supporting multidisciplinary research, training and policy development linked to disasters and climate change globally, regionally and nationally. Leading gender mainstreaming in this area is the Institute for Gender and Development Studies (IGDS). The Institute's masters and doctoral degree programs provide advanced education and research. Included in the BSc in Gender and Development is education to expose students to issues related to men and masculinities, and to mainstreaming gender in climate change and disaster risk management. Much of this work in the Caribbean has focused almost exclusively on women, hence the importance of broadening the discussion to include boys and men. Chant and Gutmann's (2000) appeal to mainstream men into gender and development work is therefore relevant given the links between gender, development and disaster risk.

In 2009, the UWI's IGDS Mona Unit conducted eight studies for UNDP's Caribbean Risk Management Initiative (CRMI) to assess the institutional capacity of national disaster management agencies in Belize, Dominica, Jamaica, Guyana and Suriname. Men and masculinities emerged as an important theme in the eight studies. They highlighted men's traditional roles as protectors and their risk-taking behaviors; men's flood-related vulnerability to contracting leptospirosis; and injury to men during hurricanes and floods as they sought to protect their houses and livestock and rescue their own and neighbors' property. The findings also revealed occupational stereotypes and a clear division of labor, with males in top leadership positions of disaster response agencies, some of whom were former military officers; women held supportive roles with limited power and authority to influence policy decisions (UNDP, 2009). Knowledge of gender and technical capacity for mainstreaming gender were limited. Policies of the agencies were "gender blind," which assumed that there is no difference between men and women despite differences in age and other socioeconomic backgrounds which would affect their respective abilities to prepare for and respond to disasters. The exception was that mothers with young babies are generally given special treatment. Agency representatives indicated a willingness to learn how to mainstream gender in their policies and programs, providing a foundation for these institutions to consider intersecting gender differences and planning more effectively for males and females.

Dunn's (2013) edited volume on *Gender, Climate Change, and Disaster Risk Management* included research and analysis by students enrolled in the UWI BSc Gender and Development course. Shaniquea Ormsby (2013) examined differentials in risk perception among Jamaican males and females and the higher risk of women and girls to gender-based violence after a hurricane. Joshauna Small (2013) examined the gendered effects of climate change on (male) livestock farmers, the majority of whom remain in place, rather than moving to a shelter, in order to protect their livestock, housing and other assets. Kevon Kerr (2013) highlighted the vulnerabilities of Jamaican fishermen livelihoods in coastal areas affected by rising sea levels, while Sheldon Gray (2013) highlighted the increased risk of Jamaican males' exposure to malaria from playing football at dusk in a suburban community. Ann Marie Virgo (2013) assessed the gender impact of droughts on male farmers in rural communities, noting how the lack of water challenged men's masculine role and identity as family providers; drought also forced migration as a survival strategy, and resulted in negative health outcomes due to reduced access to clean drinking water, to stress and respiratory illnesses, and to poor nutrition linked to food insecurity. These papers are informed as well by Dunn's earlier (2009) regional discussion of environmental justice in relation to natural disasters and climate change, examining the roles and livelihoods of men and women in Jamaica working in such key economic sectors as tourism and bauxite mining, and the need to protect workers' rights in Small Island Developing States in the Caribbean.

Frameworks for change in gender and Caribbean disaster risk management

The previous sections of this chapter demonstrated the vulnerabilities of Caribbean SIDS to climate change and natural hazards; global regional and national commitments to gender mainstreaming for gender equality and sustainable development; the need to use intersectionality to strengthen this process; and the value of using this approach which highlighted the specific vulnerabilities of groups of males, including MSM, poor homeless gay men and men with disabilities.

I conclude by suggesting a number of practical steps for disaster risk management agencies. Improving capacity to collect data disaggregated by sex and other demographic and socioeconomic characteristics is one, as this enables disaster managers to better identify and address the specific needs of males across the life cycle, as well as men with disabilities, men of different class and racial/ethnic groups, and men of different sexual identities. Agencies will also need to look at these intersecting gender patterns and how they can impact men's access to power and decision-making with the aim of more effective disaster risk management. Also important are the assets that diverse groups of men bring to policy-making and problem solving, such as local knowledge networks, resources, cultural insights and trust—all of which are strategically important to the disaster management process. More gender-related research is also needed on men, masculinity and disasters in the Caribbean. Findings can be used to develop more inclusive, targeted and effective policies and programs. UNDP should build on the CRMI Project Initiative to monitor and share these and related changes in policies, programs and practices relating to men, and masculinities.

Partnerships between disaster risk management agencies, academic institutions and MSM advocacy groups can help to harness specialized knowledge and other assets that men can provide to mitigate disaster-related risks that affect them. For example, the specialized knowledge assets and competencies of MSM, men with disabilities, poor men in rural Caribbean communities, older men in urban inner cities or working-class men employed in utility companies to provide water, electricity and sanitation and other services can be used to enhance disaster risk management practices and better protect men's lives. Partnerships with networks of specific groups of men are important, as members are likely to know and may trust each other. These social capital assets and cultural insights can be harnessed to support the development of effective early warning systems, and promote behavior change to mitigate risks.

Disaster risk management agencies can also use this knowledge to mainstream gender equality principles in institutional policies and legal frameworks; to promote policy coherence by closing the gap between current practices and global human rights commitments to integrate gender in disaster risk management; to develop relevant Gender Polices and Gender Action Plans focused on awareness building and training, as well as building staff capacity to collect and analyze data disaggregated by sex and other factors; and to use gender analysis tools to develop evidence-based policies and programs. To ensure gender equality, enhanced gender mainstreaming must be supported by intersectionality analysis with specific focus on diverse forms of masculinities.

The new knowledge from gender mainstreaming can also help to change the patriarchal culture of some disaster management agencies, structured exclusively on hegemonic masculinity. It is important that the increased focus on men should complement (not replace) existing strategies that focus on the needs of women and girls. Government agencies responsible for climate change and disaster risk management should provide an enabling environment to facilitate and strengthen multidisciplinary institutional partnerships between disaster management agencies and academic institutions, national gender machineries, agencies addressing gender-based violence, sexuality and employment, and advocacy groups supporting men's rights. The CARIMAN Initiative and Fathers' Inc. are two of several groups that can provide critical insights and expertise to share on men, masculinities and disaster management.

Caribbean institutions are demonstrating an increased awareness of the need to pursue inclusive and participatory approaches resulting in win-win outcomes for highly vulnerable men and other persons. By building on the strategic development goals in Vision 2030 Jamaica and the Post 2015 Development Agenda, climate change and disaster risk management policies and practices will by 2030 reflect greater understanding of and engagement with Caribbean men as key stakeholders.

References

Anderson, P. (2009) 2000 round of population and housing census data analysis sub project, national census report 2001, Jamaica. Available from http://www.caricomstats.org/Files/Publications/NCR%20Reports/Jamaica.pdf (accessed September 13, 2015).

Boxill, I., Galbraith, E., Mitchell, R. and Russell, R. (2012) National survey of attitudes and perceptions of Jamaicans towards same sex relationships: A follow-up study. Kingston: University of the West Indies. Available from http://www.aidsfreeworld.org/RSS/~/media/Files/Homophobia/Jamaica%20National%20Survey%20on%20Homophobia.pdf (accessed August 16, 2015).

Bradshaw, S. (2004) Socio-economic impacts of natural disasters: A gender analysis. Santiago Chile: United Nations CEPAL. Available from http://www.cepal.org/mujer/reuniones/conferencia_regional/manual.pdf/ (accessed October 17, 2015).

Bradshaw, S. (2013) *Gender development and disasters*. Cheltenham: Edward Elgar Publishing.

Bradshaw, S. (2015) Engendering development and disasters. *Disasters*, 39(1), pp. 54–75.

BRIDGE (2008) Gender and climate change: Scoping study on knowledge and gaps. London: DFID. Available from http://www.bridge.ids.ac.uk/sites/bridge.ids.ac.uk/files/reports/Climate_Change_DFID.pdf (accessed December 15, 2015).

Caribbean Risk Management Initiative (CRMI) (2009) Enhancing gender visibility in disaster risk management and climate change in the Caribbean. Barbados: UNDP. Available from http://crmi-undp.org/en/genderstudy/index.php (accessed September 12, 2015).

Chant, S. and Gutman, M. (2000) *Mainstreaming men into gender and development*. Oxford: Oxfam Working Papers.

Clatterbaugh, K. (1990) *Contemporary masculinity: Men, women and politics in modern society*. Boulder, CO: Westview Press.

Dunn, L. (2009) The gendered dimensions of environmental justice: Caribbean perspectives. In: Chioma-Steady, F. (ed.) *Environmental justice in the new millennium: Global perspectives on race ethnicity and human rights.* New York: Palgrave Macmillan, pp. 111–133.

Dunn, L. (ed.) (2013) *Gender, climate change and disaster risk management.* Working Paper 7. Kingston: Institute for Gender and Development Studies Mona Unit, the University of the West Indies and the Friedrich Ebert Stiftung.

Dunn, L. and Sutherland, V. (2009) *Gender based violence in Jamaica: A profile of victims and perpetrators.* Kingston: Friedrich Ebert Stiftung.

Government of Jamaica (2014) Disabilities Act 2014. Available from http://www.japarliament. gov.jm/attachments/341_The%20Disabilities%20bill%202014%20No.13.pdf (accessed August 14, 2015).

Gray, S. (2013) Gender socialisation and malaria risks in Portmore St. Catherine. In: Dunn, L. (ed.) *Gender, climate change and disaster risk management.* Working Paper 7. Kingston: Institute for Gender and Development Studies Mona Unit, the University of the West Indies and the Friedrich Ebert Stiftung, pp. 22–29.

Hankivsky, O. (2014) Intersectionality 101. The Institute for Intersectionality Research & Policy, SFU. Available from https://www.sfu.ca/iirp/documents/resources/101_Final. pdf (accessed September 17, 2015).

J-FLAG (2014) Annual report: Engaging communities transforming lives. Available from http://jflag.org/wp-content/uploads/2014/08/Annual-Report-JFLAG.pdf (accessed July 14, 2015).

Jonkman, S.N. and Kelman, I. (2005) An analysis of the causes and circumstances of flood disaster deaths. *Disasters,* 29(1), pp. 75–97.

Kerr, K. (2013) Climate change and natural disaster risk management in coastal areas: A gender perspective. In: Dunn, L. (ed.) *Gender, climate change and disaster risk management.* Working Paper 7. Kingston:, Institute for Gender and Development Studies Mona Unit, the University of the West Indies and the Friedrich Ebert Stiftung, pp. 44–51.

Mishra, P. (2009) Let's share the stage: Involving men in gender equality and disaster risk reduction. In: Enarson, E. and Chakrabarti, P.G.D. (eds.) *Women, gender and disaster: Global issues and initiatives.* London: Sage, pp. 29–40.

Ormsby, S. (2013) Perceptions of risks of hurricanes among Jamaican males and females. In: Dunn, L. (ed.) *Gender, climate change and disaster risk management.* Working Paper 7, Kingston: Institute for Gender and Development Studies, Mona Unit, the University of the West Indies and the Friedrich Ebert Stiftung, pp. 15–21.

Planning Institute of Jamaica (2012) Jamaica survey of living conditions 2012. Available from http://www.pioj.gov.jm/Portals/0/Social_Sector/Executive%20SummaryFinal. pdf. (accessed July 14, 2015).

Reddock, R. (ed.) (2007) *Interrogating Caribbean masculinities: Theoretical and empirical analyses.* Kingston: The University of the West Indies Press.

Small, J. (2013) The gendered effects of climate change on livestock farmers in the Caribbean. In: Dunn, L. (ed.) *Gender, climate change and disaster risk management.* Working Paper 7. Kingston: Institute for Gender and Development Studies Mona Unit, the University of the West Indies and the Friedrich Ebert Stiftung, pp. 30–34.

UNDP (2009) Making disaster risk reduction gender-sensitive: Policy and practical guidelines. Geneva: UNISDR, UNDP and IUCN. Available from http://www.preventionweb. net/files/9922_MakingDisasterRiskReductionGenderSe.pdf (accessed July 14, 2015).

Virgo, A.M. (2013) Gender impacts of droughts. In: Dunn, L. (ed.) *Gender climate change and disaster risk management.* Working Paper 7. Kingston: Institute for Gender and Development Studies Mona Unit, the University of the West Indies and the Friedrich Ebert Stiftung, pp. 66–75.

19 Men, masculinities and disaster

An action research agenda

Elaine Enarson

Dangerously increasing climate and disaster risk now leads the world's people to a "new normal" that manifestly jeopardizes lives and livelihoods across the globe. This social fact alone brings men to the forefront, given the power of men in managing the social processes through which disaster and climate risk are constructed and distributed. Yet, as noted in this book's introduction to the literature (Enarson and Pease, in this volume), and despite important new work engaging men and masculinities (Reid, 2011; Pacholok, 2013; Parkinson and Zara, 2013; Eriksen, 2014; Pease, 2014; Rumbach and Knight, 2014), we see large blank spaces and only a few fully developed corners when asking "what men do" in disasters (Enarson, 1998, 2001).

To help address this gap, I offer three main and intertwined lines of research, teased apart here for my purposes. These must be examined at different levels of analysis and in different contexts; they are not sequential and should not be tied unnecessarily to the disaster "cycle" metaphor. Community-based action research and feminist methodologies, with their shared histories and communities of practice, are especially useful for gender and disaster research (Reason and Bradbury, 2013; McCall and Peters-Guarin, 2014). Theory must always drive method, but it is essential to capture sex- and age-specific data, and allow for comparative, longitudinal, interdisciplinary and intersectional analysis. This is an action research agenda. Working with local communities affords new insight into local cultural knowledge, and working with practitioners opens the door to applying new knowledge. Leaving our respective corners for the uneasy spaces between disciplines and affinities is the most challenging yet promising pathway for those who want knowledge to guide progressive change. I hope and expect gender and disaster researchers will in future include more men with this intent. This field needs the talents of all people skilled in gender analysis and committed to action research for social change.

Male experience in disaster studies is too often confounded with human experience; for example, generalizing from male patterns of homelessness, or strategies for economic recovery based on (some) men's work lives. The disaster experiences of diverse boys and men must be understood on their own terms, though of course men's experiences cannot be divorced from women's nor from women's empowerment and disempowerment in disasters. The diverse responses of men to perceived security threats, for instance, profoundly affect the lives, livelihoods and autonomy of other men and of the women in their lives. These intersections must be reflected in the questions researchers ask and answer.

With these caveats, I would like to suggest three broad lines of inquiry: first, understanding hazards and disaster risks in the "whole community"; second, illuminating the "doing" of masculinity in the "doing" of disaster risk management; and third, bringing men's particular gender interests and practices front and center in future because, in the end, we must not simply manage disaster risk better but reduce the driving forces that increase risk and undermine resilience, including embedded inequalities. My discussion ends optimistically with some observations about how to put the new knowledge we may gain to use, working as a united and inclusive action research community.

Understanding disaster risk: practical knowledge about men and masculinities

Understanding risk as a function of hazard, vulnerability and capacity (Wisner *et al.*, 2004) is the critical first step to reducing it. We must know more about the everyday living conditions of diverse men and boys, building the fine-grained analysis necessary to inform preparedness and mitigation campaigns, training, communication and outreach, evacuation planning, shelter management, housing reconstruction, volunteer management, emergency response and relief efforts, recovery planning and all the other issues that keep disaster managers up at night. A local and inclusive lens is essential.

Hazard exposure

Just as surely as military service and armed conflict take men's lives and hyper-masculine work cultures lead them to early "manly" deaths (Schmoll, 2013), the routine livelihood, domestic and leisure activities of men and boys, as well as male-dominated organizational roles, inevitably expose them to diverse hazards. Male dominance in resource-dependent occupations and lifeline industries helps explain their disproportionate exposure to toxic hazards, illustrated recently by the male damage control teams responding to the Fukushima Daiichi disaster. Men's exposure obviously increases when military facilities are targeted in purposive attacks or explosions. Yet, the contours of daily work may also be protective; for example, keeping men on the high seas in a tsunami while women perish on the shores awaiting the catch (Oxfam International, 2005). The social fact that men are not generally primary hands-on caregivers to the ill, formally or informally, will prove protective in climate-related health emergencies. But what new threats on the horizon will endanger men and boys differently? What impacts can be anticipated on men due to crisis-driven migration, and to male-specific effects on livelihood as the planet warms (Hunger and David, 2011; Goh, 2012)? Risk management demands empirical answers.

Risk perception and tolerance

Identifying as male and demonstrating manliness is never an inherent prescription for risky behavior. But, considering risk-tolerant gender norms in male-dominated sports, business and military activities, it is a short leap to conclude that hegemonic gender norms endanger many boys and men in the face of hazards and disasters. Few researchers have examined differences among men with respect to environmental hazards (but see Becker, 2011) though particular men in risky environments evince different degrees of risk perception and tolerance (see Maier, 1997, on the Challenger accident; and Miller, 2012, for a case study of drilling and water contamination in the United States). Researchers point to the relatively benign worlds of white men as a factor in contrasting male perceptions of risk, with less privileged men more closely approximating the lower risk tolerance levels reported by women (Finucane *et al.*, 2000). From this perspective, evaluating strategies for decreasing risk tolerance among boys and men most at risk is essential public safety research.

Social vulnerabilities and capacities analysis

Widely and often justly critiqued, disaster vulnerability analysis nonetheless moved the field substantially toward structural differences and inequalities and away from the prevailing "special needs" of those "special" populations out of sync with the (default) avatar in disaster management – white, heterosexual, able-bodied, middle-class men. The vulnerability concept has greatly influenced what we know now about men in disasters. Much of this work is psychosocial, illuminating emotional harm to men in disaster contexts; but, many other questions are pressing, including around mortality, health, livelihood and housing. How are specific groups such as gay youth and seniors, transgendered men, gay men living in poverty or with disability likely to fare in different disaster contexts?

Women do not always or necessarily endure higher death rates in disasters, despite their common vulnerability in particular political-economic contexts (see Neumayer and Plümper, 2007). Evidence from the US and elsewhere suggests men die more often than women due to flooding, to take one example (Jonkman and Kelman, 2005). Giving less credence to posted warnings, overestimating their driving skill or the capacities of their autos, responsibilities toward dependents, and urgent need to protect land, livestock and livelihood surely explain this as well as (better than) a generalized critique of masculine norms. But we must learn much more about which men die more often than women, under what conditions and why. More importantly, how are men and boys in different contexts hurt or disabled, and with what long-term effects on their lives after disaster? Importantly, what long-term reproductive harms can be documented among men exposed to toxics and other threats in their roles as responders, formal or informal?

Posttraumatic stress disorder (PTSD), a concept imported into disaster studies with little critical analysis, has informed most psychosocial research with male disaster responders and survivors. Some researchers will want to explore men's mental health in disasters with more nuance. Suicidal behaviors and other self-destructive actions pertinent to men's lives in disasters are as yet poorly understood, too. What promising practices can be identified and evaluated? For instance, what gender-responsive interventions can be identified and evaluated that have targeted men living with PTSD or abusing alcohol or drugs in the wake of disaster? How do men interpret their own use of alcohol and/or drugs, with what implications for their health and the well-being of others?

Given the frequency with which disaster-related gender violence is reported, researchers will also want to identify causal factors promoting and mitigating violence in men's diverse relationships, including violence against women and same-sex partners. One small study from Texas noted the link between African-American male teenagers' exposure to a hurricane (Meshack *et al.*, 2012) and the post-disaster domestic violence they committed, clearly inviting more study. There is a great need to examine institutional violence, as well. Which men face disaster-related violence at the hands of the state, including from uniformed responders, contract workers, law enforcement and the military? How do they respond and with what consequences? What protects those men most at risk from disaster-related human rights violations, including in complex emergencies that compound natural hazards and armed conflict, create massive refugee crises and offer such opportunities for sex-specific violence perpetrated by and against male boys and men?

Men in "lifeline" transportation, telecommunications and energy industries, like the many millions whose livelihoods derive from natural resources, are hit hard by environmental disasters, including those of the present and future tied to global warming. So what delays, impedes or supports different men's economic recovery in different contexts? How do men and boys in the informal economies that dominate our planet survive and move forward after disasters, if at all? What about men in temporary encampments and in new ethnic enclaves created or increased by climate-driven displacement? What gender disparities are institutionalized in economic relief and recovery programs that jeopardize the recovery of men and male youth stigmatized as sexual minorities? The secondary risks to men and boys in the aftermath, including disaster-related trafficking of homeless boys and others into forced labor and/or sex work, demand analysis if positive interventions are to be effective.

Vulnerability analysis also takes researchers to housing and shelter. How are preparedness and mitigation furthered, if at all, in male dormitories provided by employers, rooming houses frequented by male migrants, group homes for disabled men and other male-dominated facilities? How structurally resistant are they and other state-operated physical facilities that disproportionately house low-income men of color, such as prisons, jails and halfway homes?

With respect to households and families caught up in disasters, a host of questions arise for new researchers focused on diverse men and masculinities. How do social relations evolve, including fictive kin relationships, in the homosocial worlds of men post-disaster (in work camps, on reconstruction crews far from home, in refugee camps) and with what effects on the women in their lives and others? What forms of disaster caregiving do they take up, with what consequences for family well-being? How do single fathers negotiate the demands of family and work in disasters, and long-distance fathers in transnational households? How do displaced fathers, sons and brothers stay connected, if at all? Knowing that marriage is a protective factor for men more than women (Norris *et al.*, 2002), we must also explore how this may or may not vary in men's same-sex relationships and with what significance for men's emotional and financial recovery. How does being a grandfather or being widowed matter? *Kodokushi* or "death by loneliness" is a concept employed to help explain post-disaster male mortality in Japan, clearly warranting much more investigation in other cultures and societies, as do the specific needs of men with disabilities in the aftermath, both in private residences and in shelters, temporary accommodations and long-term caregiving institutions.

Importantly, a corollary subset of questions about capacities and resources follows on these lines of research around the root causes of men's potential vulnerabilities in disasters. At a minimum, the relevant life experiences that diverse men draw upon to anticipate, prepare for, resist, cope with and recover from disasters can be systematically identified and appreciated. What skills and insights have they gained living life as men, and in what ways do diverse men help others to survive and thrive, or simply to negotiate the challenges disasters bring? What do men's friendship, leisure, workplace and kin networks bring to men in crisis, and for how long? We must also learn much more about how men catalyze the resources they control or have access to, in whose interests and with what effects. I return to this in the final section, but move first from understanding to managing disaster risk.

Managing disaster risk and managing gender

A second major line of research lies in untangling the convergent, competing and conflicting gender norms and hierarchies that affect the real-time management of disasters and disaster risk, both in discourse and in practice. What consequences follow for disaster reduction and for gender equality? These are very large research areas with significant applications in the practices of disaster risk management.

Risk assessment and information management

What data, information and knowledge inform disaster risk assessments, the building blocks of risk management? In the end, risk assessments are useful only insofar as they are timely, relevant, nuanced—and as fine-grained as possible, with disaggregated data on a small scale. How often do we achieve this, thinking about gendered disaster vulnerabilities and capacities? Do some men speak for all? Future researchers may want to assess and evaluate how or whether existing "gender-inclusive" data banks actually represent the lived realities of diverse boys and men caught up in disasters, and of transgendered persons and all others whose very identity challenges easy check-box approaches, for example in post-disaster needs assessments. Are dissenting male voices marginalized, for example, around climate change, or sexuality, or work and family conflicts in first response professions? Critically, as is so often noted, we must not assume that findings based on the male/female sex variable promote gender analysis, no matter how convenient one-dimensional data points are in electronic risk mapping.

Disaster risk communication, education and awareness campaigns

Highly gendered in image, text and messaging, and implicitly targeting men, disaster risk communications rarely take gender into consideration. With "information" the presumptive tool for survival, it is important to know more. What critical messages about risk reach specific subgroups of boys and men, with what consequence for community-wide information exchange? Access to and control over the tools of communication (new and old) are not equally distributed, so disaster communicators need to know how best to reach those most in need, boys and men among them. How can those most marginalized be reached in their everyday lives with potentially life-saving knowledge? Who has effective control over key messaging around disaster risk, and how is this determined? Researchers with expertise in communications and media studies should assess the scope and significance of disaster communication through a masculinity lens, examining such highly masculinized media as disaster video games, "Master of Disaster" board games and action-packed edutainment videos produced by tornado "storm chasers." Others might investigate how, if at all, male and female teachers strive to specifically reach young men with useful knowledge about self-protection in disasters, or evaluate the use of popular male "champions" to reach boys and men in risky environments, as in antiviolence campaigns.

Mitigation and preparedness

Informal nonstructural mitigation of known and recurring hazards is the norm at the grassroots level, where women take the lead—for example, in rainwater harvesting and household and neighborhood conservation efforts. But mitigation at large reads male: physical hazards and structural vulnerabilities are emphasized, and male-dominated mitigation strategies. Men's cultural power, including traditional Indigenous knowledge, to mitigate known hazards is often overlooked, but elite men's overall control over key resources, like male-to-male networking with governmental and financial "stakeholders," is hard to miss. We must explore these patterns. Are male-led mitigation efforts, in fact, tied empirically to any particular approach—soft or hard, high tech or low, structural or nonstructural? Is lobbying local governing agencies to reinforce seismically vulnerable school buildings, or organizing neighborhood garbage collection to clear local waterways in flood times, men's work or women's or both? European and American men, some studies find, are less supportive than women of concrete steps toward reducing emissions, though their carbon footprints are greater (McCright, 2010; Ergas and York, 2012). What does this imply for male-led climate negotiations and comparable work around reducing disasters? Without close focus on the part played by gender, along with social class and other drivers of difference, we cannot hope to challenge these and other disparities in climate mitigation and adaptation.

Similarly, men and women seem to move toward preparedness, if at all, along highly gendered pathways, with men concentrating their efforts around the exterior and in public domains. But what impact does this have on overall family and community preparedness? Neighborhood emergency preparedness efforts are increasing as emergency authorities continue to devolve responsibility and reduce government support, so we must know who is involved, how, why and with what outcomes. One dimension of preparedness is readiness (and capacity) to respond. While work and family conflict is assumed in the case of female first responders, it is not consistently seen as a concern of men, including male disaster managers, nurses, firefighters and others whose presence is assumed. Do men "just say no" to loved ones or to employers when their responder roles conflict with family, including in dual-career households and in an era of increasing opportunity for women emergency managers (Scanlon, 1998)? Further, how does parenting, under different disaster risk conditions, affect the preparedness intentions and actions of diverse groups of fathers, grandfathers and other male guardians?

Evacuation

Like risk perception, gender is a highly predictive factor in the decisions made by people of all genders about whether, how and when to evacuate. Findings from developed societies such as the United States and Australia indicate men's lower propensity to evacuate (Rosenkoetter *et al.*, 2007; Eriksen, 2014), while boys and men are better protected in rural societies with shelters that prove inaccessible to women not accompanied by male relatives. In addition to other known markers of evacuation intention and capacity, masculine norms and their effect on family decision-making must be analyzed to help planners and responders target their efforts: Which men are most likely to decline or delay evacuation, why and with what consequences for their safety and others, including other men who will attempt their last-minute rescue? When partners disagree, in same-sex relationships and traditional ones, whose voice dominates, and how might risk managers reach these evacuation resisters?

Emergency response

Autonomy, self-sufficiency, altruism, physical strength, bravery and stoicism—all are enduring tropes of manliness strongly on display in disasters, especially among the uniformed responders who stand in culturally for male heroism. Yet, men's personal accounts around heroism in crisis tell a more complicated human story. Qualitative research is replete with accounts from women of male partners and kin emotionally immobilized and unable to ask for help. We might examine how or whether diverse and gender-balanced teams of outreach relief workers make it easier—and how often nontraditional roles are open to male volunteers. From a different perspective, how do male survivors evaluate the acknowledged gender bias toward women that often exists in distribution of food vouchers? Both criticized and applauded, the practice clearly needs investigation for its unintended consequence includes increased domestic violence, some studies suggest (Crawford, 2013). Does conflict among men also increase, or does men's health and nutrition status decline, for example, among men not closely connected to women (food) through family or kin networks?

More work is essential to understand the significance of competing forms of masculinity among disaster responders such as firefighters and others whose gender subjectivities, practices, interests and strategies may divide them in the face of crisis. The highly militarized approach to disaster management (command and control) is seen to disempower women as it empowers (some) men. How does this volatile mix take shape, and with what effects for men needing and providing relief, when disasters unfold in diverse cultural and political-economic contexts? With respect to emergency mass care, we must better understand how different shelters plan (or fail to plan) for different groups of men, among them young boys whose presentation may put them at risk or men who depend on others for personal care or who arrive and leave the shelter alone and poor. Male-specific hygiene kits may be available, but are condoms (where culturally accepted) or other reproductive health services for men? Clearly, a better grasp of gender violence in shelters and camps must be a priority. Researchers working with a foot both in critical men's studies and disaster studies can flesh out the picture we now have of sexual assault and such coercive sexual practices as sexual harassment, food-for-sex and survival prostitution; effective interventions must build on knowledge of how this varies in encounters with male intimate partners and with men representing external militaries, NGOs, governments, private relief groups, contract response agencies and other disaster actors.

Reconstruction and recovery

Without attention to marginalized men, male youth and boys, including transgendered men, disaster "recovery" cannot be attained. Differences in men's housing, transportation and economic needs and resources are not simply incidental or individual but reflect patterns of power and privilege that cannot be overestimated in disaster recovery. For example, the housing needs of single fathers, widowers, male transients and migrant laborers must be understood locally when post-disaster housing initiatives are planned. What are the long-term survival strategies of male subsistence farmers or agricultural workers, if displaced first by a flood or earthquake and then by economic need—and how does this affect the women in their lives and communities? Given the increase in transnational households, dependent upon remittances from men, women or both, we must ask how dependent families survive when disaster intervenes. Men are generally highly involved in post-disaster reconstruction, in governmental and nongovernmental programs, as manual labor, as professionals in design and construction, and in the financial and regulatory institutions that direct post-disaster reconstruction. Does this matter? How can we know? Further, paid construction work is a short-term resource largely reserved for (certain groups of) men. What is the role of male-dominated unions in disaster reconstruction? New and emerging climate realities make crisis-driven male migration a more pressing issue in disaster work. We need to know how likely different groups of men are to return to their home countries and neighborhoods. Important differences in post-disaster economic opportunities will surely emerge among men in different social and geographic locations as the effects of a warming planet are further revealed—and must be well understood to develop effective interventions and support.

As mentioned earlier, disaster recovery research focused around men and masculinities should explore PTSD indicators for subgroups of men; for example, a risk factor among men is trauma from prior sexual assault (Leck *et al.*, 2006). Future research should relate more strongly to hegemonic norms of masculinity but also to men's progressive work as men against violence against women. Thoughtful critiques are needed of training and practice guidelines, with special attention to whether or how effectively these reflect knowledge about men's mental health in disasters. Monitoring and evaluation research on the ground must follow.

Gender mainstreaming in disaster management

Just as "resilient" is not an inherently positive adjective (male dominance proving very resilient in disasters), gender mainstreaming is not inherently liberatory. A skeptical stance is warranted toward the gender mainstreaming project generally and men's concerns specifically in disaster risk management (Aolain, 2011) as in development work (Cornwall *et al.*, 2011; Jauhola, 2012). Yet, it is important to examine exclusionary practices here in how differently, if at all, men are engaged in gender and disaster mainstreaming projects (Mishra, 2009). How male power regimes dominate the culture and practices of disaster management and with what consequences is an essential problem in the social production and reproduction of risk and demands attention regardless of the sex of those who walk in the disaster manager's shoes.

How does the numerical domination of men in key disaster management positions actually shape the way disaster management is done, and how does this also engage the politics of race, of sexualities, and of age and disability? We might also ask what, if any, changes follow when male disaster managers or practitioners do not, in fact, promote the highly masculinized and contested "command and control" approach; instead, they may (or may not) bring gender equality values and expertise from social work, counseling, community organizing and other fields far from the work cultures of dominant emergency management occupations. When sex ratios do shift in key positions, do changes in approach, focus, preferred technologies or other decision points necessarily shift as well? Moving beyond generalized critiques of hegemonic domination in emergency management, future researchers may find that men's voices can be heard among the many "excluded voices" (Hewitt, 1998) that yet arise in opposition to the bureaucratic administration of disasters and emergencies, especially those living lives of subordinated masculinities. Action researchers are well positioned to work with organizational critics and leaders to analyze promising and less promising practices potentially leading toward stronger and more democratic approaches to disaster management (see Parkinson *et al.*, 2015).

Reducing risk: transformative knowledge about men and masculinities

As important as applied research is, the root causes of disaster risk must be the focus of future research, recalling the words of African-American feminist activist Flo Kennedy : "If you're lying in a ditch with a truck on your ankle, you don't send somebody to the library to find out how much the truck weighs. You get the truck *off*" (quoted in Steinem, 2013). Root cause research is not an indulgence but essential in an era of increasing risk, ever more powerful private markets and weaker states, rising inequalities, global environmental and human security challenges, and resilient structures of race, class and gender domination.

"Man-made" disasters

To call disasters "man-made" is to say nothing about which men or in whose interest this transpires—but neither is it simply a footnote to reducing disaster risk. Problematizing the practice of male dominance in the culture and practices of disaster management is not a second-order question but an essential problem in the social production and reproduction of risk.

While most disaster risk analysis is stubbornly hazard-focused, blind to social organization generally and gender specifically, powerful early writing opened the door to gender, development and disaster analysis. But male elites deeply engaged in the production and distribution of risk are generally seen as acting on the basis of class and racial privilege alone, as if sex, sexuality and gender were not part of their lives. This misapprehension of capitalism and other macro-economic patterns of domination as gender-neutral (Acker, 2006) has masked gender interests in social life, and complicates our efforts to untangle the roots of disaster risk today. Racialized gender power, and not class interest alone, is in play in the wake of disasters. Risk analysis with a critical men's studies perspective substantially advances analysis of disaster capitalism. Toward this end, "studying up" (David, 2010) is a sorely needed approach, one that particularly challenges but does not preclude participatory action research.

At one level, disaster risk is manifestly constituted by men acting in their own self-interest, "behind closed doors," and in ways that exacerbate or create vulnerability for others (see Nagel, 2016, for the case of climate risk). But we must ask harder questions. Critical men's studies offers trenchant analysis of diverse men's gender subjectivities and practices in the military, firefighting and emergency medical care, in first response occupations, social welfare agencies, public administration and governance institutions, as well as in science and technology organizations and a host of other disaster-relevant research sites. How are conflicting masculinities expressed in post-disaster tension and conflict between survivors and disaster relief workers, social workers, engineers, bankers, insurance agents, politicians and other elite men in the uniform of disaster helpers? Violence and conflict in post-disaster scenarios invite research from this perspective. When "fighting" is said to break out between male survivors or between male beneficiaries and their putative "helpers," what diverse gender interests and practices guide them? How does gendered conflict among men in disaster contexts shape community recovery or vulnerability to future risk factors?

Manly technologies of risk

The "pleasure and power" (Hacker, 1989) of men and technologies invite research into gendered strategies of risk management, including structural mitigation and adaptation. The discourse of control ("man vs. nature") has long characterized dominant approaches to disaster risk. This desired male control over unruly forces is very much a part of the cultural landscape and social history in which development decisions were made that manifestly put more people at increased risk. Dams, levees, pipelines, complex integrated telecommunications networks and other enormous public works projects around the world—designed, financed, constructed and administered by different classes of men—too often increased hazards and disasters in the end. What historical lessons can be learned that bear on the future—for instance, proposed geo-engineered solutions to the threat of planetary destruction? With the entry of more diverse groups of men and of women into engineering, construction, architecture and related fields, would researchers discern any shift toward or away from the "hard" approaches to mitigation, adaptation and other complex human problems on our dynamic planet? The story of how development decisions designed the disasters of today is often told—but the story about the men behind this has not yet been heard. Massive "man-made" development projects clearly benefit women whose everyday lives are made easier, yet simultaneously endanger or disempower millions more. It is important for researchers to tease apart these privileges and dangers.

Masculine workplace cultures

Critical men's studies of highly masculinized workplaces offer disaster researchers a solid foundation for analyzing change in emergency management workplaces, including the part leadership styles play in challenging gendered occupational safety norms (Somerville and Abrahamsson, 2003). Empirical findings about men's disaster work (paid and unpaid) can support needed changes that better promote and protect men's health and well-being, including in first responder workplaces and other roles in which men are less visibly engaged. In organizations where wrong decisions may increase people's vulnerabilities to hazards and disasters, we can learn much more about the part men's workplace associations, sports teams, unions and other work-based networks play in decision-making around risk and risk reduction. This is not an arcane area of interest but germane to positive change in disaster jobs and occupations that bear on risk and risk reduction. Ethnographic studies of workplace cultures functioning as high pressure chambers of hegemonic masculinity are needed when these bear on disaster risk, for example in deep-water oil drilling, bioengineering, robotics, space exploration and global financial management. On the flip side, what organizational structures and systems best support shifts toward forms of manliness conducive to participatory, democratic and rights-based approaches to reducing disaster risk?

Men organizing to reduce disaster risk

As important as it is to delineate the mechanisms and strategies through which men's gender interests are reasserted and reclaimed in crises, students of disaster and masculinity will also want to document and analyze how men can and do work toward social justice, sustainable development and risk reduction. Never losing sight of the context-specific nature of hazards and disasters or the many voices with which men speak, we must explore men's organizing around disaster risk reduction. Despite grassroots women's activism, elite men have historically dominated mainstream environmental organizations. In what other ways do men work toward safer, more just futures—in their backyards and homes, schools and workplaces, through faith-based and political organizations, collectively, individually and as profeminist men in gender justice movements?

At a different level, and across the globe, legions of men step up to help minimize known hazards and prepare their homes and communities for the unknown. In the aftermath, they often lead formal mitigation and "build back better" resilience efforts, bringing a lifetime of male training, skills, networks and resources to bear. Male volunteers dominate, for self-evident reasons, in the drawn-out process of physically reconstructing stricken communities. Military veterans apply their hard-won skills in the aftermath, for example, through the Team Rubicon network in the US (http://www.teamrubiconusa.org/). In emergencies, local talk show hosts, generally male, are trusted by men whose lives mirror their own (Laska, 2013). These and similar initiatives that insightfully engage men with other men in mitigating hazards, reducing vulnerabilities and building capacities should be studied in the interest of public safety.

Concluding reflections

The urgency of understanding the dynamics of a warming planet and its effects on people and place inspires a new generation of researchers from which disaster and gender work can only benefit. The questions I have raised here will be answered differently, and lead to new questions and answers. As action researchers, we insist on using time and resources in ways that engage us equitably and effectively with our research partners, always with an eye on the prize: How will this new knowledge promote change? Crises reveal the fundamental structures of social life, including how disasters increase and/or decrease male power, but we must have eyes to see and the desire to look. Examining gender relations in degraded physical and social spaces and specific environmental and cultural contexts will bring a needed place-based dimension to masculinity studies (Connell and Messerschmidt, 2005). Equally, when disaster studies takes on board what has been learned about men resisting and redirecting power and privilege (Pease, 2000, 2010), new partners in reducing risk are revealed and new voices for progressive change. It is past time to consider how the geographies of risk and of masculinities are joined and with what consequences for reducing disaster and climate risk and moving toward gender equality and women's empowerment. This is neither women's nor men's work, nor does it belong solely in the domain of disaster and climate research.

But men's engagement as producers and users of new knowledge is an essential part of the solution we must find together.

References

Acker, J. (2006) *Class questions: Feminist answers*. Lanham, MD: Rowman & Littlefield.
Aolain, F. (2011) Women, vulnerability, and humanitarian emergencies. *Michigan Journal of Gender and Law*, 18(1), pp. 1–23.
Becker, P. (2011) Whose risks? Gender and the ranking of hazards. *Disaster Prevention and Management: An International Journal*, 20(4), pp. 423–433.
Connell, R.W. and Messerschmidt, J. (2005) Hegemonic masculinity: Rethinking the concept. *Gender & Society*, 19(6), pp. 829–859.
Cornwall, A., Edström, J. and Greig, A. (eds.) (2011) *Men and development: Politicizing masculinities*. London: Zed.
Crawford, E. (2013) 2010 Haitian earthquake: Investigation into the impact of gender stereotypes on the emergency response. MA. Oxford Brookes University. Available from http://architecture.brookes.ac.uk/research/cendep/dissertations/Emma-Crawford-dissertation.pdf (accessed February 16, 2016).
David, E. (2010) "Studying up" on women and disaster: An elite sustained women's group following Hurricane Katrina. *International Journal of Mass Emergencies and Disasters*, 28(2), pp. 246–269.
Enarson, E. (1998) Through women's eyes: A gendered research agenda for disaster social science. *Disasters*, 22(2), pp. 157–173.
Enarson, E. (2001) What women do: Gendered labor in the Red River Valley flood. *Environmental Hazards*, 3, pp. 1–18.
Ergas, C. and York, R. (2012) Women's status and carbon dioxide emissions: A quantitative cross national analysis. *Social Science Research*, 41(4), pp. 965–976.
Eriksen, C. (2014) *Gender and wildfire: Landscapes of uncertainty*. New York: Routledge.
Finucane, M., Slovic, P., Mertz, C.K., Flynn, J. and Satterfield, T. (2000) Gender, race, and perceived risk: The "white male" effect. *Health, Risk & Society*, 2(2), pp. 159–172.
Goh, A. (2012) A literature review of the gender-differentiated impacts of climate change on women's and men's assets and well-being in developing countries. International Food Policy Research Institute. CAPRi Working Paper 106. Available from http://cdm15738.contentdm.oclc.org/utils/getfile/collection/p15738coll2/id/127247/filename/127458.pdf (accessed November 2, 2015).
Hacker, S. (1989) *Pleasure, power, and technology: Some tales of gender, engineering, and the cooperative workplace*. Boston: Unwin Hyman.
Hewitt, K. (1998) Excluded perspectives in the social construction of disaster. In: Quarantelli, E.L. (ed.) *What is a disaster?* New York: Routledge, pp. 75–92.
Hunter, L. and David, E. (2011) Displacement, climate change, and gender. In: Piguet, É., Pécoud, A. and Guchteneire, P. (eds.) *Climate change and migration*. New York: Cambridge University Press, pp. 306–330.
Jauhola, M. (2012) *Post-tsunami reconstruction in Indonesia: Negotiating normativity through gender mainstreaming initiatives in Aceh*. New York: Routledge.
Jonkman, M. and Kelman, I. (2005) An analysis of the causes and circumstances of flood disaster deaths. *Disasters*, 29(1), pp. 75–97.
Laska, S. (2013) "Citizen" responders: Ordinary men making extraordinary moves through radio programming. Available from http://scholarworks.uno.edu/cgi/viewcontent.cgi?article=1028&context=chart_pubs (accessed February 2, 2016).

Leck, P., Difede, J., Patt, I., Giosan, C. and Szkodny, L. (2006) Incidence of male child-hood sexual abuse and psychological sequelae in disaster workers exposed to a terrorist attack. *International Journal of Emergency Mental Health*, 8(4), pp. 267–274.

Maier, M. (1997) Gender equity, organizational transformation and Challenger. *Journal of Business Ethics*, 16(9), pp. 943–962.

McCall, M. and Peters-Guarin, G. (2014) Participatory action research and disaster risk. In: Wisner, B., Gallard, J.C. and Kelman, I. (eds.) *The Routledge handbook of hazards and disaster risk reduction and management*. London: Routledge, pp. 772–785.

McCright, A. (2010) The effects of gender on climate change knowledge and concern in the American public. *Population and Environment*, 32(1), pp. 66–87.

Meshack, A., Peters, R., Amos, C., Johnson, R., Hill, M. and Essien, J. (2012) The relationship between Hurricane Ike residency damage or destruction and intimate partner violence among African American male youth. *Texas Public Health Journal*, 64(4), pp. 30–33.

Miller, L. (2012) Women and risk: Commercial wastewater injection wells and gendered perceptions of risk. In: Measham, T. and Lockie, S. (eds.) *Risk and social theory in environmental management*. Collingwood, VIC: CSIRO Publishing, pp. 130–146.

Mishra, P. (2009) Let's share the stage: Inclusion of men in gender risk reduction. In: Enarson, E. and Chakrabarti, P.G.D. (eds.) *Women, gender and disaster: Global issues and initiatives*. Delhi: Sage, pp. 29–39.

Nagel, J. (2016) *Gender and climate change: Impacts, science, policy*. New York: Routledge.

Neumayer, E. and Plümper, T. (2007) The gendered nature of natural disasters: The impact of catastrophic events on the gender gap in life expectancy, 1981–2002. *Annals of the Association of American Geographers*, 97(3), pp. 551–566.

Norris, F., Friedman, M., Watson, P., Byrne, C.M., Diaz, E. and Kaniasty, K. (2002) 60,000 disaster victims speak: Parts 1 and 2. *Psychiatry*, 65(3), pp. 207–260.

Oxfam International (2005) The tsunami's impact on women. Oxfam Briefing Note. Available from http://policy-practice.oxfam.org.uk/publications/the-tsunamis-impact-on-women-115038 (accessed November 2, 2015).

Pacholok, S. (2013) *Masculinities in crisis: Gender at work in the wake of disaster*. Toronto: University of Toronto Press.

Parkinson, D. and Zara, C. (2013) The hidden disaster: Violence in the aftermath of natural disaster. *The Australian Journal of Emergency Management*, 28(2), pp. 28–35.

Parkinson, D., Zara, C. and Davie, S. (2015) Victoria's gender and disaster task force. *The Australian Journal of Emergency Management*, 30(4), pp. 26–29.

Pease, B. (2000) *Recreating men: Postmodern masculinity politics*. London: Sage.

Pease, B. (2010) *Undoing privilege: Unearned advantage in a divided world*. London: Zed Books.

Pease, B. (2014) Hegemonic masculinity and the gendering of men in disaster management: Implications for social work education. *Advances in Social Work & Welfare Education*, 6(2), pp. 60–72.

Reason, P. and Bradbury, H. (eds.) (2013) *Handbook of action research: Participative inquiry and practice*. London: Sage Publications.

Reid, M. (2011) A disaster on top of a disaster: How gender, race, and class shaped the housing experiences of displaced hurricane Katrina survivors. PhD. University of Texas.

Rosenkoetter, M., Covan, E.K., Bunting, S., Cobb, B. and Fugate-Whitlock, E. (2007) Disaster evacuation: An exploratory study of older men and women in Georgia and North Carolina. *Journal of Gerontological Nursing*, 3(12), pp. 46–54.

Rumbach, J. and Knight, K. (2014) Sexual and gender minorities in humanitarian emergencies. In: Roeder, L. (ed.) *Issues of gender and sexual orientation in humanitarian emergencies*. New York: Springer, pp. 33–74.

Scanlon, J. (1998) The perspective of gender: A missing element in disaster response. In: Enarson, E. and Morrow, B.H. (eds.) *The gendered terrain of disaster*. Westport, CT: Greenwood, pp. 45–52.

Schmoll, B. (2013) Masculine and dead in the mining community: The gendering of death and the Monongah mine explosion of 1907. *Journal of Appalachian Studies*, 19(1/2), pp. 27–45.

Somerville, M. and Abrahamsson, L. (2003) Trainers and learners constructing a community of practice: Masculine work cultures and learning safety in the mining industry. *Studies in the Education of Adults*, 35(1), pp. 19–34.

Steinem, G. (2011) The verbal karate of Florence R. Kennedy, Esq. *Ms Magazine* blog, summer. Available from http://www.msmagazine.com/summer2011/verbalkarate.asp (accessed February 16, 2016).

Wisner, B., Cannon, T., Blaikie, P. and Davis, I. (2004) *At risk: Natural hazards, people's vulnerability and disasters*. London: Routledge.

Afterword

Raewyn Connell

PROFESSOR EMERITA, DEPARTMENT OF SOCIOLOGY, UNIVERSITY OF SYDNEY

The nearest I have come to a major disaster was five years ago in Yogyakarta. Gadjah Mada University was holding a conference when Mt Merapi blew up, 30 km to the north. The international visitors woke to find the city covered with volcanic ash, the airport closed, trains disrupted and refugees coming in. The university closed and turned into a refugee reception center. Quick work by our hosts got the visitors shipped out by road via the south. I arrived in Jakarta ill after an improvised cross-country trip, but I got off lightly: 350 people died, and 350,000 were evacuated from their homes.

Like many disasters this eruption seemed to have a beginning and an end; but as this book forcefully shows, there is also a long aftermath, a downstream that doesn't end quickly. And I've come to realize something else this book shows—that some disasters will come to me wherever I retreat; and the news from others will keep coming too.

Research on disasters, and official policy about disaster prevention and response, have gradually come to include issues about gender. That change has drawn on a generation's worth of research on gender, a field transformed by the impact of feminism in the 1970s and still developing in new ways. Gender research has become global—it's active in Gadjah Mada University for instance—but has had difficulty about how to be global. Though Europe and North America produce the bulk of research and theory, it simply doesn't work to impose Global North theoretical models on the enormous diversity and different historical experience of the postcolonial world. Intellectuals in the South do produce powerful ideas about gender, but their ideas usually don't circulate widely (Connell, 2014a).

Research on masculinities and men as gendered social beings has flourished as a field of gender analysis for the last 30 years, and has internationalized. Masculinity research too has had difficulty coming to terms with the postcolonial terrain, though the issue is increasingly recognized (de Jesus 2011; Connell 2014b). It's very good to see research and experience from 11 different countries in this book, with attention to questions of Indigeneity, development and postcolonial conflict.

A notable feature of the study of masculinities was the early development of applied research, taking the findings of the new field to practical problems of social life. This includes work in education, in health, in counseling and therapy, in youth work, in violence prevention and in quite a range of social policy areas. As the editors observe, *Men, Masculinities and Disaster* is not the first application of a critical masculinities approach to issues that arise from disasters. But it is the first wide-ranging and systematic treatment of these issues. The book makes a strong contribution not only to disaster research but to studies of gender and masculinities too.

One of the strengths of the book is the range of methods the contributors use, including participant observation, qualitative interviewing with samples of men, focus groups, autoethnography, case studies and—a valuable method that's too rare in masculinities research—interviews with women about men.

The wealth of data these methods generate does more than illuminate different ways men construct masculinity in a context of disaster. They also tell us something about disasters themselves, as social processes. The lessons include the way disasters challenge and erode the state, its agencies and its legitimacy; the greatly varying stresses that disasters put on regional communities; and the ways disasters undermine trust and cooperation in marriages. These studies of men and masculinities under stress have lessons for our thinking about masculinities in general. Two seem to me particularly important.

First, a striking feature of the book is the prominence of institutional and collective masculinities in these studies. It's common in everyday language, and in conventional social science, to understand masculinity as a property of an individual, for instance as a personal identity. But in these studies of disaster, we see how specific masculinities are embedded in, and asserted or defended by, a social organization or collectivity. Cases include the state and its agencies; an occupational group (e.g. the striking case of firefighters, discussed in several chapters); a union or a social movement. There's an important lesson here for strategies of change: that shifts in individual identity are not enough.

Second, the close-focus studies bring out the importance of masculinity *dynamics*, i.e. processes of situational engagement and change. We miss far too much if we assume masculinities are defined by stable social norms. Certainly we see in many of these studies appeal to conventions and norms, sometimes rigid belief in norms. But these norms are often under challenge—from women, from dissident men, from public policy and from the impact of the disasters. What we see here is the *negotiation* of masculinities under pressure, sometimes in a rapidly changing situation. That negotiation may result in violence, in relationship breakdowns, in reassertions of an old hegemony—and all those cases are discussed here. But it may also result in changed gender practices, and questioning of the previously unquestioned—and we see those cases, too.

The book also brings new perspectives to agendas for change in masculinity. We get pointers to sites where change can happen, and moments when it might: occupational mentoring, for instance is worth thinking about in many contexts; the moments when social movements engage with sexist media, or masculinized institutions and governments; educational work in development contexts; traumatic loss and the process of mourning. There are risks in all of these situations, as there are risks in any agenda of change in the gender order. A sympathetic focus on men can turn into a defense of men's interests and a portrayal of men as gender victims—that has already happened in other arenas! A critique that attributes oppressive behavior simply to "traditional masculinity" or "traditional norms" ignores the diversity of traditions. The multiplicity of masculinities and their traditions is a resource for equity, as Kopano Ratele (2013) has strongly argued in South Africa.

There are also risks in typical ways we research masculinities. A common procedure, feeding a set of interview transcripts into a qualitative-analysis computer program that indexes the set as one gigantic text, tends to produce an overhomogenized picture of norms and discourses, and frequently misses the situational dynamics of gender. Many discussions of masculinity fall back into a kind of gender-role theory that, focusing on conformity to norms, misses power, economics and institutional processes. It is important to recognize that "gender norms" include norms *for equality*, as well as norms for hierarchy (Pearse and Connell, 2015).

Yet these risks are worth taking because the stakes are so high. We need experiment and imagination, in addition to well-grounded knowledge. We need bravery, among men and among women, to achieve change. And we need men as well as women to take responsibility for change toward more equal and peaceful gender relations. The studies of disasters and disaster response in this book show some groups of men taking that responsibility, others refusing it. Change in the gender order is a political arena, not a mechanical one. But that also means it is an arena of hope.

References

Connell, R. (2014a) Rethinking gender from the South. *Feminist Studies*, 40(3), pp. 518–539.
Connell, R. (2014b) Margin becoming center: For a world-centered rethinking of masculinities. *NORMA: International Journal for Masculinity Studies*, 9(4), pp. 217–231.
de Jesus, D.S.V. (2011) Bravos novos mundos: Uma leitura pós-colonialista sobre masculinidades ocidentais. [Brave new worlds: A post-colonialist reading of Western masculinities]. *Estudos Feministas*, 19(1), pp. 125–139.
Pearse, R. and Connell, R. (2015) Gender norms and the economy: Insights from social research. *Feminist Economics*. November. Available from http://dx.doi.org/10.1080/13 545701.2015.1078485 (accessed February 16, 2016).
Ratele, K. (2013) Masculinities without tradition. *Politikon: South African Journal of Political Studies*, 40(1), pp. 133–156.

Index